# Springer Tracts in Modern Physics 115

# Springer Tracts in Modern Physics

---

* denotes a volume which contains a Classified Index starting from Volume 36

Bene Poelsema   George Comsa

# Scattering of Thermal Energy Atoms

## from Disordered Surfaces

**With 74 Figures**

Springer-Verlag Berlin Heidelberg GmbH

Dr. Bene Poelsema
Professor Dr. George Comsa

Institut für Grenzflächenforschung und Vakuumphysik
KFA Jülich, Postfach 19 13
D-5170 Jülich, Fed. Rep. of Germany

---

*Manuscripts for publication should be addressed to:*
Gerhard Höhler
Institut für Theoretische Kernphysik der Universität Karlsruhe
Postfach 69 80, D-7500 Karlsruhe 1, Fed. Rep. of Germany

*Proofs and all correspondence concerning papers in the process of publication should be addressed to:*
Ernst A. Niekisch
Haubourdinstraße 6, D-5170 Jülich 1, Fed. Rep. of Germany

---

ISBN 978-3-662-15091-7      ISBN 978-3-540-45983-5 (eBook)
DOI 10.1007/978-3-540-45983-5

Library of Congress Cataloging-in-Publication Data. Poelsema, Bene, 1946–. Scattering of thermal energy atoms from disordered surfaces. (Springer tracts in modern physics; 115) Bibliography: p. Includes index. 1. Surfaces (Physics) 2. Scattering (Physics) 3. Helium–Thermal properties. I. Comsa, George, 1931–. II. Title. III. Series. QC1.S797 vol. 115 [QC173.4.S94] 530 s 89-5895 [530.4]

# Preface

The progress of surface science is still dominated by the continuous invention and implementation of new investigation methods. The permanent need for new, more efficient methods is due to the complexity of surface phenomena and the ever increasing variety and importance of their technical applications. Another reason is that one particular method supplies information only about a limited aspect of a given phenomenon. Two or more complementary methods are in general necessary for a detailed description and especially for a deep understanding of a given aspect.

Surface disorder has been suspected for a long time of playing a decisive role – positive or negative – in a number of surface processes, many of them of particular technological significance. One of the significant findings of the famous scanning tunneling microscope has been to demonstrate in a direct way how much disorder is present on surfaces thought to be "almost perfect". The method, based on thermal energy atom scattering (TEAS), that we are reviewing here, appears to complement in an ideal way scanning tunneling microscopy in the investigation of disordered surfaces. The unique atomic resolution of tunneling microscopy confines its reasonable application to microscopic surface areas; the method has inherent difficulties with dynamical processes and delicate objects (e.g. weakly bound adsorbates). These are precisely the strengths of TEAS. Besides information on the structure of some surface objects, TEAS supplies information on average properties like lateral distribution of adsorbates and defects, and on the relative location of adsorbates and defects as well as on coadsorbed species. Extremely low adsorbate coverages and defect densities can be detected straightforwardly.

The low energy of the thermal beams and the inertness of He-atoms makes the probing process fully non-destructive: neither the most unstable surface objects nor the most delicate surface processes are significantly influenced by the measurement. This allows the continuous in situ monitoring of a variety of processes such as island formation, two dimensional condensation, defect formation and sputtering by ion bombardment, layer growth by vapor deposition, and many others. Another characteristic of TEAS due also to the low energy of the incident atoms is its absolute surface sensitivity; the scattering takes place in the electron sea outside the outermost surface layer.

The various applications of TEAS are based mainly on diffuse scattering and interference. We have tried to make a presentation accessible to as large a readership as possible. The simplicity of the ideas on which TEAS is based were particularly helpful in this respect.

We hope that this review will inspire the reader to share our enthusiasm for the He-scattering approach to surfaces; and this not only for the quality of the supplied information, but also for the straightforward way to obtain it, with a minimum of theoretical complication.

Most of the work presented has been performed in the authors' laboratory. We wish to thank all who have contributed to experiments, discussions of the results and new ideas during a longer or shorter period of time: Steven Bernasek, Larry Brown, Rudolf David, Klaus Kern, Klaus Lenz, Günter Mechtersheimer, Robert Palmer, Laurens Verheij, Peter Zeppenfeld and Siebe de Zwart. Our own investigations have benefitted from the skillful technical assistance of Herbert Kleingans. They would have been hardly possible without the "ideal" crystal surfaces prepared by Udo Linke. We have learned much in various stages of the work from discussions with Hans Bonzel, Mark Cardillo, Vittorio Celli, Gerhard Ertl, Benny Gerber, Harald Ibach, Jean Lapujoulade, Sieghard Lehwald, and Andrea Levi; to Dick Manson and John Parmeter we are grateful in addition for the critical reading of the manuscript. Last but not least we thank Maria Kober for the careful typing of the manuscript.

Jülich, June 1989                           *Bene Poelsema · George Comsa*

# Contents

# 1. Introduction

The progress in experimental science comes from observing and exploring regularities. This is certainly a triviality; it offers, however, an explanation why systems and phenomena, which are characterized by a high degree of order, are more frequently subject of scientific investigation. Confining the discussion to surfaces, much more is known about the properties of ordered surfaces than of non-ordered ones. We know for instance the structure of a large number of surfaces, but we do not know even the nature of defects which are present in addition to some step rows on a well prepared surface of a metal monocrystal.

Many of the methods developed and used so far have accentuated this imbalance. Staying with the example above, diffraction methods, in particular low energy electron diffraction (LEED), which have contributed decisively to decipher the nature of ordered structures, have been obviously less efficient with disorder. LEED patterns of apparently excellent quality (bright and sharp spots) may be obtained from surfaces with a substantial amount of disorder [Houston, Park, 1970; 1971]. They give valuable information on the structure of the well-ordered fraction of the whole area, strengthen the self-confidence of the experimentalist but hardly give information on the disordered fraction of the surface. The information on disorder contained in the level and structure of the background is not easily accessible. From this point of view, diffraction has a certain resemblance with physicists: they both emphasize the regular ordered aspects of objects and phenomena, while repressing dubious, disordered ones somewhere in the background. The importance of disorder in surface science can, however, hardly be overestimated. Indeed, any real surface, even the best prepared surface of a monocrystal, contains a certain amount of disorder; and minute disorder like residual steps or impurities may influence decisively surface processes like reactivity or crystal growth. Moreover adsorption, desorption, surface reaction and diffusion, phase transitions, crystal growth, sputtering and roughening intrinsically involve disorder. A detailed knowledge of the nature, distribution and dynamics of surface disorder is thus ultimately as necessary as that of surface order.

The invention of the scanning tunneling microscope [Binnig, Rohrer, Gerber, Weibel, 1982] and especially its enthousiastic use in many laboratories to "look" at surfaces with atomic resolution is of overhelming significance for the investigation of surface disorder, see e.g., [Garcia, 1987; Feenstra, 1988]. On the one hand it made clear to anyone how imperfect "perfect" surfaces are. On the other hand, the microscope is able to uncover directly the nature of the individual constituents of surface disorder. Like any other method, scanning tunneling

1

microscopy has its limitations and has to be combined with other methods. In view of the atomic resolution and the real space imaging it is hardly possible to obtain information on the actual distribution of disorder over macroscopic areas. The monitoring of the dynamical processes enumerated above, in particular those involving mobile constituents, is also hardly possible. The approach based on the scattering of thermal He beams on disordered surfaces discussed in this review appears to ideally complement the scanning tunnel microscope. Indeed, both the investigation of the lateral distribution of adsorbates and defects and the monitoring of the various dynamical surface processes are the outstanding features of the thermal He beam scattering approach.

In spite of the fact that thermal He diffraction on surfaces has been demonstrated in the late twenties [Stern, 1929] only a couple of years after electron diffraction, the development of thermal energy He-atom scattering (TEAS) as a surface investigation method has been extremely slow for decades. The main reason was the lack of adequate He-beam sources. The effusive (Knudsen) cells used for many years are able to supply only a low intensity, non-monochromatic (Maxwellian) He beam. It was only the advent of nozzle-beams [Kantrowitz, Grey, 1951; Becker, Bier, 1954; Mason, Williams, 1972; Boato, Cantini, Mattera, 1976; Horne, Miller, 1976; Cardillo, Becker, 1978], which led to the breakthrough of TEAS during the seventies. Both intensity *and* monochromaticity were simultaneously improved by orders of magnitude, a progress only comparable with that brought about by lasers. The first broad range of application of the new sources has been the investigation of ordered surface structures, in particular of adlayer structures by He diffraction, see e.g., [Engel, Rieder, 1982]. Outstanding features of thermal He scattering like exclusive sensitivity to the outermost surface layer, sensitivity for hydrogen layers and non-destructiveness were crucial in the successful deciphering of delicate structures in this classical area of low energy electron diffraction (LEED). A decisive contribution to the widespread acceptance of He scattering as an efficient investigation instrument has been the application of inelastic He scattering to the investigation of surface dynamics. The very high energy resolution at large momentum exchange, which can be attained due to the favourable energy to momentum ratio of heavy probe particles (neutron, He atoms) led to the first measurement of surface-phonon dispersion curves [Brusdeylins, Doak, Toennies, 1980; Toennies, 1984]. The contribution of He-inelastic scattering to the understanding of the surface lattice dynamics during the last decade is remarkable.

The use of He scattering as an instrument for the investigation of disordered surfaces, which will be reviewed here, is — besides the use of He diffraction for ordered structures and of inelastic He scattering for surface dynamics — the third broad application area of He beams for surface exploration. In addition to the distinctive features of He scattering mentioned above, this application takes advantage of a unique property of thermal He scattering from individual adatoms and defects: the very large cross-section for diffuse scattering (linear dimensions $\geq 10\,\text{Å}$). This unusual sensitivity of He scattering for the presence of adatoms and defects has been considered for a long time as a nuisance, and a serious handicap for the instrumental use of He scattering. The large size of the cross-section of admolecules for thermal He beams has been determined already by Smith and Merril [1970]. However, it was only after the cross-sections have been studied systematically and their nature uncovered [Poelsema, De Zwart, Comsa, 1982; 1983], that their consequent use for surface investigations became possible.

After a short discussion (Chap. 2) of some experimental aspects, Chap. 3 is dedicated to the He scattering from adsorbates and impurities, in particular to the definition, nature and properties of the cross-section for diffuse He scattering. Based upon the large size of these cross-sections, a new, very fruitful procedure — the "overlap approach" — has been developed. This procedure which supplies unique information on the lateral distribution of adsorbates, defects and their various combinations will be exposed in the first part of Chap. 4. The application of this procedure to investigate the lateral distribution of adsorbates and defects in homogeneous and heterogeneous combinations will be illustrated in Chaps. 4 and 5, respectively. Highlights will be: island formation, admolecule migration and the creation of defects by ion sputtering; the nature of the interaction (attractive vs repulsive) between the surface objects (admolecules and defects) will be emphasized. Chapter 6 is dedicated to the investigation of randomly stepped surfaces. Here again the sensitivity of He scattering to characterize the perfection of nominally flat surfaces (down to step densities of $\sim 0.001$) and to follow in detail the surface morphology during ion sputtering at various temperatures will be demonstrated. In Chap. 7 the capabilities of TEAS to analyse various aspects of surface roughening, a subject of intense current interest will be discussed. Finally, in Chap. 8 we consider complex phenomena like the kinetics of adsorption and desorption and measurements under quasi-equilibrium conditions. The capability of TEAS as a very sensitive and reproducible probe of surface coverage (down to $\approx 0.001$) will become apparent.

We hope that this review will bring the reader to share with us the enthousiasm for the He scattering approach to the surfaces; and this not only for the quality of the supplied information, but especially for the straightforward way to obtain it, with a minimum of theoretical complication.

# 2. Experimental

In this chapter we emphasize the main features of the experimental conditions required for conducting TEAS experiments as discussed in this work. It appears that these novel applications of TEAS, described below, by no means require the sophistication of the very high resolution He-scattering instruments inherent to inelastic scattering experiments. A rather conventional set-up is perfectly suited to do the job. Neither very highly monochromatic He beams nor He partial pressures at the detector of $10^{-15}$ mbar are necessary. It is only the partial pressure of reactive gases in the vicinity of the sample surface which has to be very low ($\sim 10^{-11}$ mbar); the angular spread of the beams has, in addition, to be confined within adequate but by no means excessive limits.

## 2.1 The Thermal Energy Helium Beam: Production and Detection

The classic He scattering experiments of the late 1920s and early 1930s were carried out using effusive sources. These beams have a low intensity and a poor monochromaticity (Maxwell distribution). In the early 1970s, these drawbacks were overcome by the use of nozzle beams [Kantrowitz, Grey, 1951; Becker, Bier, 1954; Mason, Williams, 1972; Boato, Cantini, Mattera, 1976; Horne, Miller, 1976; Cardillo, Becker, 1978]. Simultaneously both the intensity and the monochromaticity were enhanced substantially (see Fig. 2.1).

The monochromaticity plays a key part in diffraction experiments for structure investigation as well as in inelastic scattering experiments for the determination of phonon modes. The concept of transfer width has been introduced in order to characterize the resolution of a diffraction instrument. The transfer width can be associated with a length at the surface, which would lead to the instrumental width of the diffraction peaks. It represents the linear scale at which information on the surface periodicity can be obtained directly (without deconvolution). The transfer width is given by (see, e.g. [Comsa, 1979])

$$ w = \lambda / [(\Delta_\vartheta \cdot \vartheta_f)^2 \cos^2 \vartheta_f + (\sin \vartheta_i - \sin \vartheta_f)^2 \overline{(\Delta E)^2} / E^2]^{1/2} \quad , \tag{2.1} $$

where $\vartheta_i$ and $\vartheta_f$ denote respectively the angle of incidence and the exit angle of a beam with wavelength $\lambda$; $\Delta_\vartheta$ indicates the angular spread determined by source and detector, while $\overline{(\Delta E)^2}$ represents the energy spread. Only the angu-

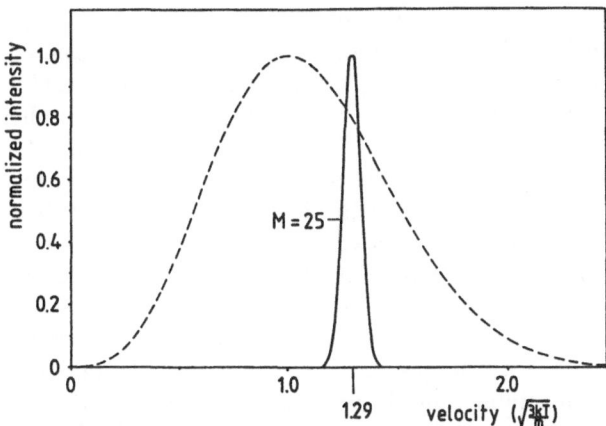

**Fig. 2.1.** Normalized velocity distributions of the flux of an effusive He beam (*dashed curve*) and of a He nozzle beam with Mach number $M = 25$ (*solid curve*) in units of the most probable flux velocity in an effusive beam

lar spread appears to be of importance for specular reflection $(\vartheta_i = \vartheta_f)$ as in most of the TEAS-applications discussed subsequently. The transfer width as defined in expression (2.1), including a certain contribution of the energy spread, plays a rôle in the interference experiments during ion bombardment induced surface damaging and during crystal growth (Chap. 6), in surface "roughening" investigations (Chap. 7), and in the hydrogen adsorption experiments described in Chap. 8.

The diffuse scattering experiments discussed in this work do not require an extremely high angular and energy resolution. Only the absolute determination of the total cross-section for diffuse scattering requires (Chap. 3) a high angular resolution. According to criteria given by Pauly and Toennies [1956], the angular resolution of the experimental set-up presented below permits a determination of the total cross-section for diffuse scattering accurate within 5%. This figure is consistent with recent results by Verheij, Lux, Poelsema and Comsa [1987].

Even under moderate expansion conditions (nozzle stagnation pressure of a few bars), the beam energy is determined uniquely by the nozzle temperature $T_0$. For a monatomic gas the beam energy amounts to $E = 2.5\,kT_0$. Accordingly, energy variations by one order of magnitude can be achieved through nozzle temperature control. An additional extension of the operational energy range is provided by beam-seeding, i.e., by mixing He with other gases. The energy of a particle with mass $m_1$, mixed in a ratio of $f/(1 - f)$ with another species of mass $m_2$, is given by

$$E_1 = E\,m_1/[fm_1 + (1 - f)m_2] \quad . \tag{2.2}$$

Beam-seeding enables the variation of the beam energy by roughly another order of magnitude.

The detection of thermal He atoms is relatively inefficient, the involved ionization process being the bottle-neck. An efficiency of about $10^{-5}$ cannot be

5

surpassed so far without affecting the linearity of the detection and without causing uncontrollable time delays. Thus the helium signal can be increased only by an enhancement of the primary intensity (high stagnation pressures and/or large nozzle diameter). This, in turn, is limited by the He gas load (typically $0.1-1$ mbar l/s) flowing into the system, which imposes high demands on the pumping system. The He partial pressure in the main chamber is kept within the $10^{-9}$ mbar range by differential pumping of the He beam line and efficient pumping in the main chamber. The latter is of crucial importance in order to maintain the base pressure of the reactive gases in the main chamber, such as $H_2$ and CO, below $10^{-11}$ mbar, and thus to allow for typical measuring periods of $15-30$ minutes. This strict requirement is necessitated by the extreme sensitivity of the thermal He beam to the presence of adsorbates. The inherent He load precludes the use of ion pumps in He beam systems. Indeed, the He gas discharge in these pumps releases earlier gettered reactive gases leading to disturbing memory effects.

Quadrupole mass filters are usually used to allow mass selective detection of the density of scattered particles. In TEAS experiments of the type presented here, the incident beam is mechanically modulated in order to permit phase-sensitive detection for signal background discrimination. The signal may be analyzed in either the analogue mode (current measurements) or in the pulse counting mode.


## 2.2  The Experimental Set-Up


The UHV-TEAS apparatus, used in most of the subsequently discussed experiments, is schematically shown in Fig. 2.2. For a complete definition of the scattering geometry the polar and the azimuthal angles of both the incident and the scattered beams can be varied independently. For this purpose, both the sample manipulator and the detector are rotatable about two axes. This set-up does not allow for differential pumping of the detector. Therefore the dynamical range of the detector, defined as the intensity range in which measurements can be carried out with time constants $\tau < 3$ s, is relatively small in the present case: only signals down to about $10^{-3}$ of the incident intensity $I_{00}$ can be measured with sufficient accuracy. A further extension of the dynamical range can be achieved only by differential pumping of the detector. Dynamical ranges of $10^{-6}$, which are required in inelastic scattering experiments, are usually accompanied by a substantial reduction of the degrees of freedom in the scattering geometry.

The gas load produced by the He source is taken care of by two differential pumping stages. The FWHM of the angular spread of the primary beam is 0.2°. The distance between detector and sample can be varied in situ between 100 and 400 mm. Correspondingly, the solid angle of acceptance in the detector is varied between 0.4° and 0.15° [Poelsema, Mechtersheimer, Comsa, 1981a].

The apparatus is equiped with the usual Auger Electron Spectroscopy (AES) and Low Energy Electron Diffraction (LEED) facilities. In addition, an ion gun is

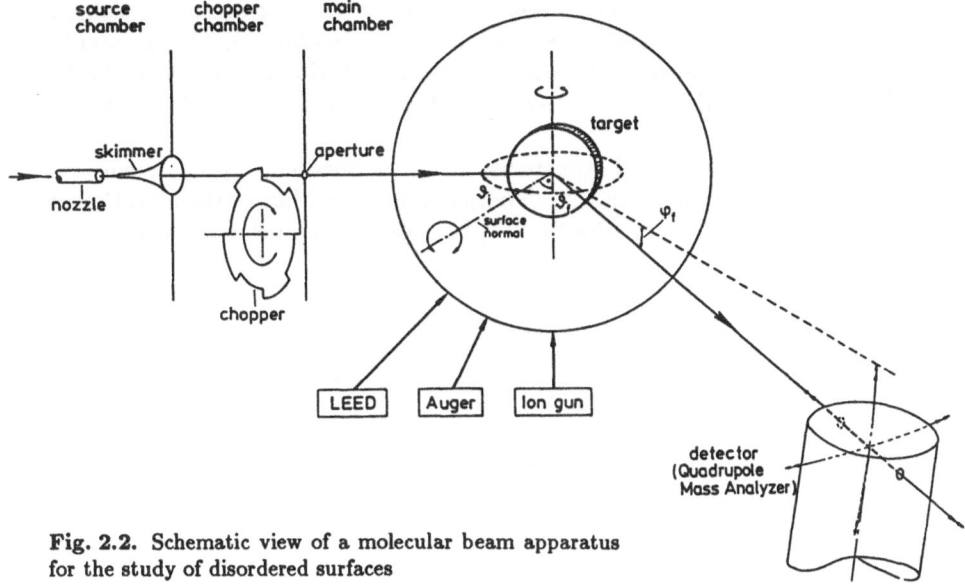

**Fig. 2.2.** Schematic view of a molecular beam apparatus for the study of disordered surfaces

mounted for sample cleaning purposes and for defined sputtering investigations (see Chaps. 6 and 8) as well as a Kelvin probe for work-function measurements.

## 2.3  Sample Preparation

Highly precise orientation and preparation of the crystal surface are a prerequisite for conducting well-defined surface experiments. In short, the procedure goes as follows: as a first step, the crystal is mounted in a special holder. Then the crystal is aligned along the wanted orientation by means of Laue diffraction; the obtained accuracy is better than 0.1°. Subsequently, the sample is cut to the final size (a disk of about 2 mm thickness) by means of spark erosion. After that the crystal is carried through a number of mechanical and (moderate) electro-polishing treatments.

Throughout the preparation procedure the sample's orientation is checked frequently by means of Laue diffraction, see e.g.[Linke, Poelsema, 1985]. It proved to be of crucial importance for obtaining well-defined surfaces that the crystals remain mounted in the same holder throughout the complete aligning and preparation procedure.

In spite of the numerous controls, the ultimate characterization of the prepared surface is only possible under UHV conditions. For this purpose, the sample is cleaned by various chemical and physical recipes, and subsequently annealed at high temperatures (ca. 90% of the melting temperature). The exact procedure depends again on the nature of the crystal; however, it always involves ion bombardment and annealing. The Pt(111) sample, used in most experiments discussed in this review, has been exposed to repeated cycles of cleaning (heating in oxygen, ion bombardment and annealing). As shown in Chap. 6, TEAS provides a unique method for the characterization of the orientation and in particular, of the defect and impurity concentrations. The procedure is discussed in detail in Chap. 6 (see also Sect. 8.4).

It has been demonstrated repeatedly that it is possible to prepare surfaces with step densities smaller than $10^{-3}$ [Poelsema, Palmer, Mechtersheimer, Comsa, 1982]. This figure implies for the Pt(111) surface average terrace widths larger than 2000 Å.

# 3.  The Scattering Mechanism

The scattering of thermal He atoms from solid surfaces is extremely sensitive to the presence of adsorbates. Although this pronounced sensitivity has repeatedly been observed [Smith, Merrill, 1970; Engel, 1978], its cause remained unresolved for a long time. The decay of the coherently scattered He intensity due to the presence of adsorbates has been quantified by introducing the concept of a cross-section for diffuse scattering [Poelsema, De Zwart, Comsa, 1982; 1983]. The strong attenuation of the He beam, already occurring at low adsorbate concentrations, corresponds to cross-sections of about 100 Å$^2$ per adparticle. In comparison to the mutual dimensions of these adparticles in their condensed state, the size of these cross-sections for diffuse scattering is certainly unexpectedly large. It is the aim of this chapter to present the definition of the cross-section, the experimental cross-section, the scattering mechanism responsible for the unusual size of the cross section, the angular dependence of the cross-section, and finally, the dependence of the He reflectivity on adsorbate coverage.

However, before doing so, we will discuss in the first two sections of this chapter the scattering process from a semi-classical point of view. This will define the framework for the subsequent considerations.

## 3.1   The He–Substrate Interaction Potential

At large separations the interaction between a He atom and the substrate is attractive. This is due to quantum fluctuations in the charge density distribution, which lead to dipole-dipole interactions between the incident He atom and the substrate atoms. These forces, often called dispersion forces, are always active on the atomic level. Decaying as $z^{-3}$, where $z$ is the distance between the He atom and the surface, these forces have a long range. At small distances, the overlap of electron clouds of the He atoms with the outer (valence) electrons of the surface atoms causes a dominant repulsive interaction. The repulsive component of the interaction potential is short-ranged. Figure 3.1a shows as an example the He-Au (jellium) potential calculated by Zaremba and Kohn [1976].

The classic He diffraction experiments of Estermann and Stern [1930] were performed on alkali-halide surfaces (e.g. LiF). Ever since then, these surfaces have played an important role in atom diffraction experiments. At alkali-halide surfaces the valence electrons are strongly localized. This fact has two important implications: (1) The equipotential-surface, as probed by the thermal He beam,

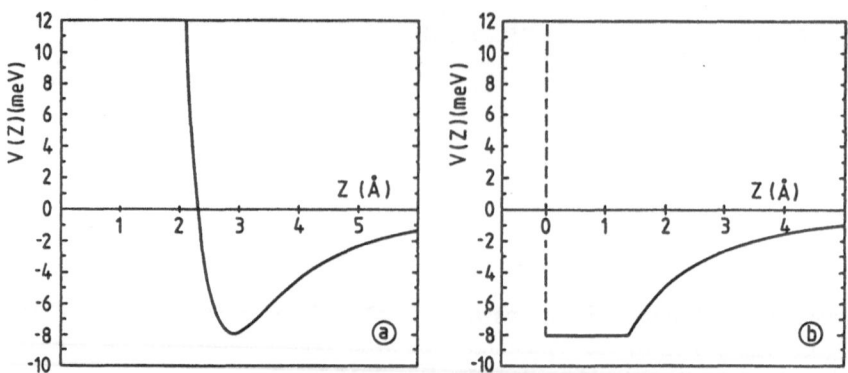

**Fig. 3.1a,b.** Potential for helium-substrate interaction: **(a)** calculated for Au jellium [Zaremba and Kohn, 1976], **(b)** in the hard-wall approximation (see text)

is strongly corrugated. This periodic corrugation is closely related to the atomic cores, enabling a direct determination of the surface structure; (2) The repulsive wall of the interaction potential is extremely steep. Therefore, the He diffraction can be described well in terms of the hard-wall model [Garcia, Ibañez, Solana, Cabrera, 1976; Garcia, 1976]. Within this model the slope of the repulsive potential is assumed to be infinite (see Fig. 3.1b); at variance with the repulsive potential, the attractive potential is laterally averaged, i.e., it is assumed to be non-corrugated. The role of the attractive interaction is thus confined to refraction effects. The introduction of a one-dimensional potential wall with depth $D$ leads to a modification of the energy and of the angle of incidence (Beeby correction [1971]). The He atoms hit the repulsive wall with an energy $E'$ and at an angle of incidence $\vartheta_i'$, which differ from the corresponding values at large separation, $E$ and $\vartheta_i$, respectively, according to

$$\sin \vartheta_i' = \sin \vartheta_i \cdot (1 + D/E)^{-1/2} \tag{3.1}$$

and

$$E' = E + D \quad . \tag{3.2}$$

This model has been applied successfully in structure investigations of alkali-halide surfaces, e.g. [Frankl et al., 1978; Garcia, Celli, Goodman, 1979; Celli, Garcia, Hutchison, 1979]. Within the same model, additional detailed information on the laterally averaged shape of the attractive potential has been obtained from selective adsorption experiments.

In contrast to the rich He-diffraction patterns obtained from alkali-halide surfaces, the patterns resulting from close packed metal surfaces are disappointing. In addition to an intense specular beam, only extremely weak higher order beams show up; the first order peaks are already $3-4$ orders of magnitude less intense than the specular one [Boato, Cantini, Tatarek, 1976; Horne, Yerkes, Miller, 1980]. The rich information accessible on alkali-halide surfaces would thus seem to be inaccessible in the case of the more frequently investigated metal surfaces. The reason for the extreme simplicity of the He diffraction pattern from close packed metal surfaces, i.e., for the very weak corrugation of the repulsive potential, has been understood only relatively late: the repulsive potential originates from the interaction of the electron cloud of the incident atom with the largely delocalized conduction electrons of the metal. The classical turning point of the thermal He atom, which is related to very small electron densities ($\approx 10^{-5}\,\text{Å}^{-3}$) of the electron cloud, is located far away ($3 - 4\,\text{Å}$) from the outermost atomic cores [Harris, Liebsch, 1982; Esbjerg, Norskov, 1980; Hamann, 1981]. Consequently, the corrugation amplitude of the repulsive equipotential surface is extremely small ($\leq 0.02\,\text{Å}$).

Also in contrast with the He/LiF system, the probed corrugation amplitudes strongly depend on the scattering conditions (atom mass, energy and angle of incidence) [Rieder, 1982]. This feature is attributed to a finite slope of the repulsive potential. Therefore, new concepts had to be developed for the analysis of scattering data from metal surfaces [Armand, Manson, 1979]. Liebsch, Harris, Salanon, Lapujoulade [1982] succeeded in describing satisfactorily experimental He-Cu data in a broad primary energy range, based on first-principle calculations of the interaction potential. As before, however, the attractive part of the interaction was treated as laterally averaged, i.e., unstructured.

The absence of structure in the He diffraction patterns of close packed metal surfaces can also be taken advantage of: it renders TEAS an extremely simple and particularly transparent analytical tool for obtaining information on a remarkably wide range of surface features. By using a TEAS instrument with a moderate dynamical range of about $10^{-3}$ (see Chap. 2), the higher order ($n \geq 1$) diffraction peaks merge into the background. The diffraction pattern merely consists of the specular He beam [Boato, Cantini, Tatarek, 1976; Horne, Yerkes, Miller, 1980]. Therefore, close packed metal surfaces (like Pt(111) in most of the examples given below) are "felt" by thermal He atoms as almost perfect mirrors. This substantially simplifies the whole problem, the experimental results, their discussion and the extraction of information. Note, however, that the novel results presented below are by no means restricted to close packed metal surfaces. The application of TEAS to more strongly corrugated surfaces, i.e., in the presence of higher order diffraction beams, is always possible. In order to simplify the discussion we will in the following (except when otherwise stated) refer only to the zero order specular beam; the conclusions hold for any other coherent (higher order diffraction) beam, if present. Likewise for simplicity, we will consider that all experimental results are obtained with an instrument with moderate dynamical range, i.e., all peak intensities smaller than $10^{-3}$ of the intensity of the specular beam will be taken to have zero intensity, being lost in the background.

## 3.2  The He–Adsorbate Interaction Potential

The presence of an isolated adsorbate is certainly expected to notably modify the extremely smooth repulsive and attractive interaction potential. As discussed in the preceding section, the repulsive He-Pt potential is determined by the electron cloud spilling over the surface. The presence of an adsorbate disturbs more or less locally this electron density distribution. Therefore, the thermal He atoms experience the presence of adsorbates as a local modification (hill or hollow) of the repulsive potential. In addition, the He atom "feels" the attractive He–adsorbate dispersion forces, resulting in a pronounced modification of the shape of the attractive potential in the vicinity of the adparticle. Consequently, the presence of randomly distributed adsorbates enhances diffuse scattering, resulting in a strong decay of the specular intensity. The dispersion forces, having a much longer range than the repulsive forces, dominate the contributions to diffuse scattering in particular at lower adsorbate coverages. In case the adsorbates build an ordered layer, they impose a periodic structure on the interaction potential, giving rise to He diffraction. As a consequence the coherent beams, including the specular He beam, may recover at least partially their intensity at larger adsorbate coverages. In this process the repulsive part of the potential plays the major part.

The influence of adsorbates on the scattering of thermal He via the modification of the repulsive and attractive potential around the adparticles has been discussed so far assuming that both the substrate surface and the adlayer atoms are rigid. With this assumption the diffuse scattering is obviously purely elastic. When the surface and adlayer atoms are allowed to vibrate, a fraction of the He beam undergoes inelastic interactions. Due to the momentum exchanged in the inelastic interactions, the He atoms are scattered into a wide solid angle and the specular beam is correspondingly attenuated. This attenuation, which is obviously surface temperature ($T_s$) dependent, is actually effective even for the clean surface: at $T_s \neq 0$ the intensity of the specular beam is always smaller than that of the incident beam, even for a perfect mirror-like close packed metal surface. This has been rationalized in terms of a Debye–Waller behaviour; e.g., the Debye temperature of the Pt(111) surface obtained from two types of He scattering experiments is $\theta_D = 231 \pm 1 \, \mathrm{K}$. In the presence of adsorbates, the probability of inelastic events is substantially increased. The thermal He atom is able to interact with both the vibrational modes of the admolecules in the binding substrate potential and the internal modes of the admolecules. This interaction is often very efficient, leading to very low "Debye temperatures" of the adlayer. For instance, the reflectivity (including the sum of all coherently scattered He atoms) of a saturated CO layer on Pt(111) is reduced even at $T_s = 80 \, \mathrm{K}$ to less than about $10^{-3}$, mainly by inelastic events.

The whole discussion of the effect of disordered surfaces on He scattering, as well as the various applications of TEAS to be presented below, is based on the attenuation of the specular He beam. The wide solid angle scattering, as a result of inelastic interactions, can thus also be classified among the other processes leading to the specular beam attenuation, under the general concept of diffuse scattering. This brings about a conceptual simplification and is in line with the possibility to use for all TEAS applications presented here an instrument with moderate dynamical range, unable to distinguish between elastic and inelastic diffuse scattering. The contribution of inelastic scattering to the cross section for diffuse scattering from isolated adsorbates is of the same order as that of the perturbation of the repulsive potential induced by the adsorbate. Indeed, an inelastic interaction with an adsorbate occurs through a collision with the *repulsive* potential. Since the size of the perturbation of the repulsive potential is typically of the order of a unit cell, the very large cross-section of isolated adsorbates for diffuse thermal He scattering is still dominated by the long range attractive potential.

Formally one may distinguish three contributions to the specular intensity from a partially covered surface:

$$\frac{I}{I_0} = f_1\left(\frac{A_S}{A_0}\right) + 2\cos\varphi \cdot f_2\left(\frac{A_S}{A_0}, \frac{A_A}{A_0}\right) + f_3\left(\frac{A_A}{A_0}\right) \quad . \tag{3.3}$$

The first and the last term represent scattering contributions from the uncovered parts of the surface and from the adsorbate covered parts, respectively. The second term describes the interference between these contributions (with phase difference $\varphi$). $I_0$ is the specular intensity at zero coverage ($\Theta = 0$). In this semiclassical concept the $A$'s represent the scattering amplitudes in the specular direction: $A_0$ for the clean surface, while $A_S$ and $A_A$ apply at finite coverage ($\Theta \neq 0$) and denote the scattering amplitude from, respectively, uncovered and adsorbate covered parts of the surface. The exact form of the functions $f_i$ ($i = 1, 2, 3$) depends on the dimensions of the adsorbate structures as well as on instrumental properties. For an ideal instrument, i.e., infinitely small angular and energy spread, or adsorbate structure dimensions well below the transfer width of the instrument, (3.3) can be written as

$$\frac{I}{I_0} = \left(\frac{A_S}{A_0}\right)^2 + 2\cos\varphi \cdot \frac{A_S}{A_0}\frac{A_A}{A_0} + \left(\frac{A_A}{A_0}\right)^2 \quad . \tag{3.3'}$$

The vibrational amplitudes of the adsorbates being in general larger than those of the substrate atoms, $A_A$ decreases with increasing temperature more strongly than $A_0$ and $A_S$. Therefore, both the second and the third term in expression (3.3) are temperature dependent. The first term only depends on temperature when the vibrational amplitudes of the uncovered substrate atoms are affected by the presence of adsorbates. Note that (3.3') reduces to the first term provided the He specific reflectivity of the adsorbed layer itself is negligibly small: $A_A \approx 0$. As will be shown below, this holds, e.g., for CO/Pt(111). The other extreme case would be the system Pt/Pt(111): after depositing a complete Pt monolayer the initial situation is restored; with $A_A = A_0$ at $\Theta = 1$.

## 3.3  The Mechanism for Diffuse Scattering from Adsorbates

This section treats in detail the diffuse scattering of thermal atoms from adsorbates. The specific contributions of attractive and repulsive forces to the total cross-section are emphasized. The adsorption system CO/Pt(111) is considered first. Figure 3.2 shows a CO adsorption curve, defined as the relative He specular peak height as a function of CO exposure. Obviously, the diffuse scattering from the adsorbed CO molecules leads to a dramatic attenuation of the specular beam. At larger coverages the He reflectivity becomes "zero" ($I/I_0 \approx 10^{-3}$) and remains at this level up to CO saturation coverage. This indicates that the CO covered parts of the surface do not contribute to the specular intensity ($A_A \approx 0$, see (3.3)). The CO molecules act as perfect diffuse scatterers. The CO/Pt(111) system is thus well-suited for investigating the diffuse scattering of thermal energy atoms.

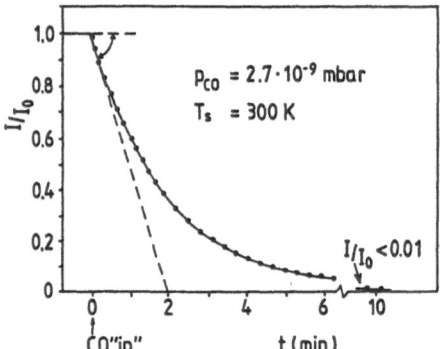

Fig. 3.2. Relative He specular peak height scattered from Pt(111) as a function of time during CO exposure at $p_{CO} = 2.7 \times 10^{-9}$ mbar and $T_s = 300$ K; $E_{He} = 63$ meV and $\vartheta_i = 40°$

Let us check whether the presence of adsorbates, which substantially attenuates the specular beam, affects also the profile of that beam. Such an experimental check is shown in Fig. 3.3, where the specular beam profile, scattered from a clean Pt(111) surface and from the Pt(111) surface partially covered with CO and H, is plotted. The partial coverages correspond to an attenuation of the specular peak height of 74% and 68%, respectively. The peak heights of the three plots have been normalized to unity. The three profiles are indistinguishable within the accuracy of the moderate dynamical range of the apparatus and also indistinguishable from the profile of the direct beam [Poelsema, Palmer, Mechtersheimer, Comsa, 1982]. This fact is of practical importance: since the peak profiles remain unchanged, the peak heights can be taken as a direct measure of the peak intensity.

In view of the discussion above, the diffuse and coherent scattering are complementary phenomena; this implies that the attenuation of the relative specular peak height, $1 - I/I_0$, is a direct measure of the probability for diffuse scattering. In the limit of high adsorbate dilution, the relative peak height varies linearly

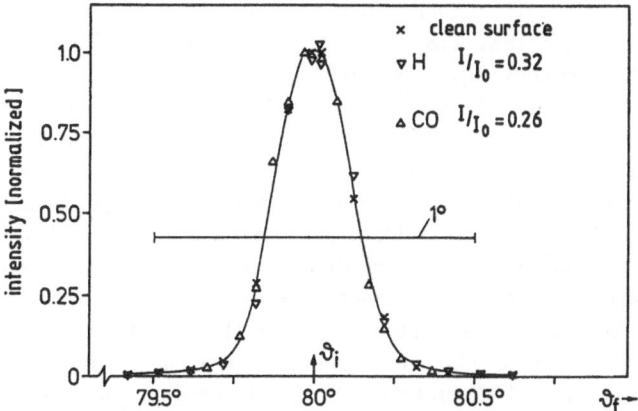

**Fig. 3.3.** Normalized profile of the He specular beam from a clean ($\times$) and an adsorbate covered Pt(111) surface; $E_{He} = 63$ meV, $\vartheta_i = 80°$ and $T_s = 300$ K. The results in the presence of adsorbates, H($\nabla$) and CO($\triangle$), were taken for $I/I_0 = 0.32$ and $I/I_0 = 0.26$, respectively

with coverage. The total cross-section for diffuse scattering $\Sigma$ can thus be defined by the expression

$$1 - I/I_0 = n\Sigma = \Theta n_s \Sigma \quad , \tag{3.4}$$

where $n$ and $n_s$ denote, respectively, the number of adsorbate and substrate atoms per unit area. In words, (3.4) ascribes an area $\Sigma$ to each individual ad-molecule; this area acts as a diffusively scattering patch on the "mirror-like" close packed metal substrate. Definition (3.4) is meaningful only as long as the "patches" do not overlap; the condition for this to occur is that the coverage is very small *and* the adparticles stay isolated (no adsorbate island formation). In more general terms, definition (3.4) can be rewritten as [Poelsema, De Zwart, Comsa, 1982; 1983]

$$\Sigma = -\frac{1}{n_s} \cdot \frac{d(I/I_0)}{d\Theta}\bigg|_{\Theta=0} \quad . \tag{3.5}$$

Therefore, in principle, the total cross-section for diffuse scattering can be derived from the initial slope of adsorption curves like the one in Fig. 3.2, as long as no island formation takes place in the considered coverage range. Note however, that while in the adsorption curve $I/I_0$ is measured as a function of the exposure $\varepsilon$, in (3.5) the dependence on coverage $\Theta$ is needed. Equation (3.5) can be written as

$$\Sigma = -\frac{1}{n_s} \cdot \frac{d(I/I_0)}{d\varepsilon} \cdot \frac{d\varepsilon}{d\Theta}\bigg|_{\Theta=0} \quad , \tag{3.5'}$$

with $(d\Theta/d\varepsilon)|_{\Theta=0}$ defining the initial sticking probability $s_0$. Thus the value of $s_0$ has to be known from an independent measurement. In the case of CO/Pt(111), measurements in several laboratories converge to a value $s_0 = 0.84$ at $T_s = 300$ K [Campbell, Ertl, Kuipers, Segner, 1981; Lin, Somorjai, 1981; Steininger, Lehwald, Ibach, 1982], which appears to apply up to $\Theta = 0.2$.

Proceeding as indicated above, the initial slope of the adsorption curve in Fig. 3.2 yields $\Sigma = 123 \, \text{Å}^2$ (corresponding to 63 meV He atoms incident at $\vartheta_i = 40°$). Cross-sections of similar size have been obtained for NO, Xe, and $O_2$ on various Pt and Ni surfaces [Wilsch, Rieder, 1983; Ibañez, Garcia, Cabrera, 1982; Ibañez, Garcia, Rojo, 1983; Poelsema, 1983]. The size of these cross-sections is at least one order of magnitude larger than the cross-sections which have been encountered before in particle-surface interactions, including the area taken by an adparticle in a two-dimensional island. The reaction of the "surface community" was correspondingly skeptical. This has certainly not been counteracted efficiently by the results of the first theoretical attempts to explain the experimental findings. These attempts were based on the assumption that the diffuse scattering was exclusively due to repulsive forces, i.e., only to the deformation of the electronic cloud induced by the presence of the adsorbates. The conclusion was that the admolecule induces a perturbation of the potential having an unusual shape and size: a Gaussian surface slipped over a cylinder with a diameter larger than 10 Å [Ibañez, Garcia, Rojo, 1983]. Although the tenacious sticking to the idea of the overwhelming role of the repulsive forces can be explained by its success in describing thermal He scattering from alkali-halide surfaces, it is, however, still surprising in view of the obvious experimental evidence (Fig. 3.4 below) for the striking similarity between the scattering from admolecules and from the same molecules in the gas phase, where the role of the attractive dispersion forces is known to be dominant. The dispersion forces between atoms and molecules are always present; why should they be switched off simply because the molecule is adsorbed? Of course, if the molecules are adsorbed in a dense 2D(imensional)-adlayer the dispersion forces — still present — are less important in the scattering process because, due to their long range, they are more efficiently averaged than the repulsive forces.

However, the cross-section for diffuse scattering is defined for isolated admolecules where, like in the gas phase, the long range dispersion forces are dominant. The gas phase scattering is well described by a Lennard–Jones–(12,6) potential [Butz, Feltgen, Pauly, Vehmeyer, 1971]:

$$V(r) = V_m \left[ \left( \frac{R}{R_m} \right)^{12} - 2 \left( \frac{R}{R_m} \right)^6 \right] \tag{3.6}$$

where $V_m$ is the depth of the potential at $R = R_m$.

While the "6-term" is an accurate description of the attractive dispersion forces, the "12-term" is only one of the handiest terms which have been chosen to describe the repulsive forces.

The size of the total cross-section for diffuse scattering in the gas phase is dominated by small angle deflections caused by long range dispersion forces. The analogy between diffuse scattering from adsorbed and free molecules was suggested first by the similarity of the cross-section sizes (e.g. adsorbed CO: $\Sigma_{CO}^{He} =$

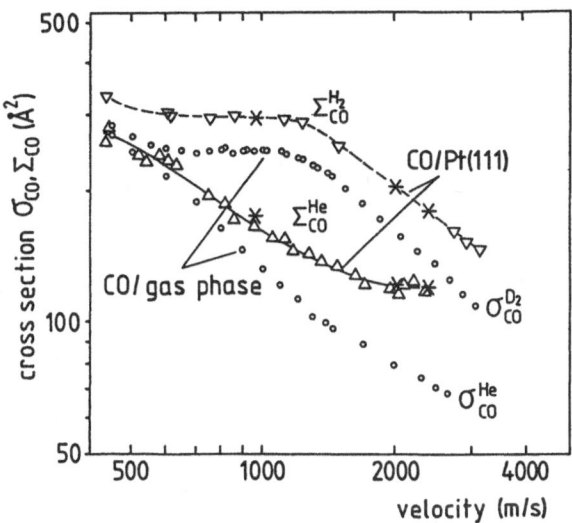

**Fig.3.4.** Measured total cross-sections for diffuse scattering on CO as a function of the velocity of the probing He atoms and $H_2$ ($D_2$) molecules. The data were taken for CO adsorbed on Pt(111) ($\Sigma_{CO}$) (probing beam incident at $\vartheta_i = 40°$) [Poelsema, De Zwart, Comsa, 1982; 1983] and for gas phase CO ($\sigma_{CO}$) [Butz, Feltgen, Pauly, Vehmeyer, 1971]

123 Å$^2$ and free CO: $\sigma_{CO}^{He} = 90$ Å$^2$ for a 63 meV He beam incident at $\vartheta_i = 40°$). The result of the conclusive experiment [Poelsema, De Zwart, Comsa, 1982; 1983], demonstrating the identity of the scattering mechanisms, is shown in Fig. 3.4. The total cross-section for diffuse scattering from adsorbed CO (on Pt(111)) has been measured as a function of the velocity of the probing incident particles (He and $H_2$). The similarity to the corresponding gas phase data, also plotted in Fig. 3.4, is obvious. The conclusion inferred by the experimentalists was that, as in gas phase scattering, the total cross-section for diffuse scattering from adparticles is determined by long range dispersion forces. The extreme sensitivity of He scattering from impurities has thus been ascribed to the adsorbate induced deformation of the *attractive* potential.

The similarity between gas - and adsorbed phase scattering can be easily understood intuitively by looking at the schematic drawings in Fig. 3.5, which show He scattering from (a) gas-phase CO and (b) adsorbed CO. The thick solid lines represent the loci of the turning points of 60 meV He-atom trajectories due to the interaction with the repulsive potential. The dashed lines are the intersection of the cross-sections for diffuse scattering (ascribed to the attractive potential) with the plane of the drawing; these cross-sections are taken as spheres centered at the CO molecules. The center of the ad-CO lies within 1 Å (for He energies in the range 10 – 100 meV) on the "mirror-plane" of the specularly reflecting Pt(111) surface. The scattering cross-section of the ad-CO is thus in good approximation a full hemisphere. The He-trajectories drawn in Fig. 3.5 show that the Pt-"mirror-plane" transforms the "surface scattering" practically into the "gas phase scattering". Indeed, trajectories Ib impinging outside the attractive hemisphere

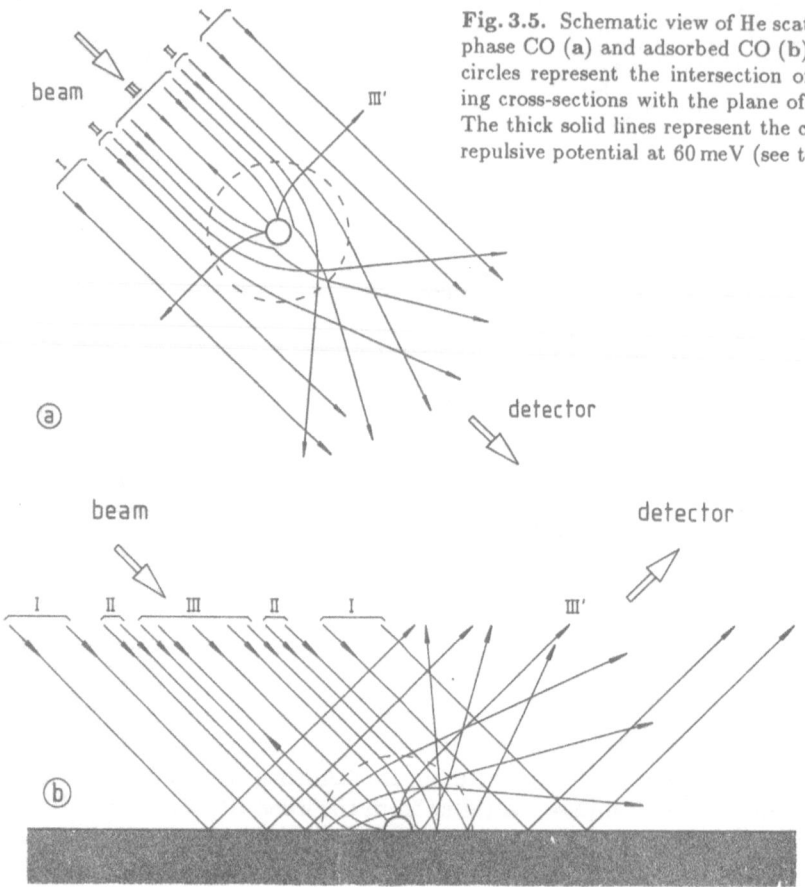

Fig. 3.5. Schematic view of He scattering on gas phase CO (a) and adsorbed CO (b). The dashed circles represent the intersection of the scattering cross-sections with the plane of the drawing. The thick solid lines represent the contour of the repulsive potential at 60 meV (see text)

(in b) result in the specular beam, which is equivalent (up to the Debye–Waller attenuation) to the direct beam Ia, which passes by the "attractive" sphere (in a). Likewise the trajectories IIb, which enter the "attractive" hemisphere, but still impinge on the "mirror", are largely equivalent to IIa, which traverse the attractive sphere but miss the repulsive one. Finally, the trajectories IIIa and IIIb, which hit the repulsive potential, are completely equivalent. Of course, all of these scattering events are equivalent up to Debye–Waller attenuations and to modifications of the repulsive electron density distribution around the adparticle and of the polarizability of the adparticle due to the presence of the substrate (see below). This, however, does not modify qualitatively the similarity of the two scattering processes. Note that the particular trajectories III', which are the result of specular reflection from the repulsive potential of the molecule, have different fates in the surface — and the gas phase case, respectively: trajectory III'b enters the detector, in contrast to III'a, which does not. It will be shown below that in general, e.g., for CO on Pt(111), this probability is vanishingly small because of Debye–Waller effects. In the case of the much stiffer H adatom on Pt(111), however, such contributions cannot be neglected (see discussion in Chap. 8).

This line of thought has been pursued quantitatively by Jónsson and coworkers [Jónsson, Weare, Levi, 1984a; 1984b; Jónsson, 1986]. The repulsive part of the He-surface potential was modelled by means of a hard wall. The long-range attractive part of the He-surface potential was accounted for by refraction effects (the so-called Beeby correction [1971], see (3.1) and (3.2). The effective potential well depth was taken to be 2.3 meV [Poelsema, Palmer, De Zwart, Comsa, 1983]. The scattering from the adsorbates was calculated with close coupling techniques. The final result is shown in Fig. 3.6, together with the experimental data. First, calculations have been performed with the best-fit parameters $V_m$ and $R_m$ (see (3.6)) [Jónsson, Weare, Levi, 1984a; 1984b; Jónsson, 1986] obtained from experimental gas phase scattering data [Butz, Feltgen, Pauly, Vehmeyer, 1971]: $R_m = 3.5$ Å, $V_m = 2.37$ meV for He and $R_m = 3.48$ Å, $V_m = 5.74$ meV for $D_2$. The calculations describe the experimental adsorbed phase data reasonably well (not shown). The excellent fit shown in Fig. 3.6 (solid curves) is obtained after a modification of the potential parameters accounting for both an enhanced polarizability of the ad-CO (a factor of $\sim 2$) and a slight enhancement (15 %) of the electron density of adsorbed CO [Jónsson, 1986]. The enhanced polarizability is in reasonable agreement with infrared measurements [Krebs, Lüth, 1977] and theoretical results [Bagus, Hermann, 1986]. For comparison, Fig. 3.6 also shows the calculated cross-section for a purely repulsive potential (dashed and dash-dotted curve). In this case the adsorbates are represented only by a hard hemisphere. This model accounts well for scattering through large angles [Lahee, Manson, Toennies, Wöll, 1986]. However, as shown by Fig. 3.6, the description of

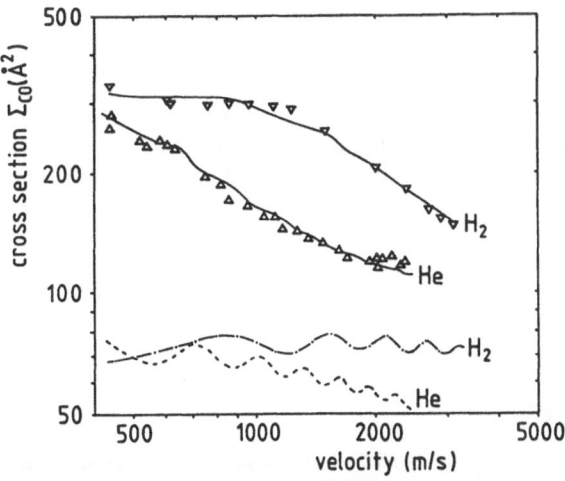

**Fig. 3.6.** Total cross-sections for diffuse scattering on adsorbed CO as a function the velocity of the probing He and $H_2$ particles. The experimental data for He($\triangle$) and $H_2$($\nabla$)were taken for CO/Pt(111). The solid curves are calculated with a gas-phase-like potential [Jónsson, Weare, Levi, 1984a; 1984b; Jónsson, 1986]. The dashed-dotted ($H_2$) and dashed (He) curves are calculated with a purely repulsive potential

the total cross-section is completely inadequate. Therefore, the following picture, consistent with gas phase scattering, emerges: the total cross-section for diffuse scattering is determined by small angle deflections, caused mainly by the always present long range dispersion forces. Large angle deflections are dominated by the repulsive part of the interaction potential. Exact calculations in the sudden approximation, as carried out recently by Yinnon et al. [1988], confirm this over-all picture. Further support for the above proposed interpretation is yielded by theoretical analyses by Gumhalter, Liu [1984a; 1984b]; Liu, Gumhalter [1986] as well as by Bosanac and Sunjic [1985].

## 3.4   The Angular Dependence of the Cross-Section

In general, the total cross-section for diffuse scattering of He atoms from ad-sorbates, as defined in (3.5), is dependent on the angle of incidence $\vartheta_i$. Figure 3.7 exemplifies this for CO/Pt(111) [Poelsema, Palmer, De Zwart, Comsa, 1983], probed with He atoms of velocity $v_{He} = 2760\,\mathrm{m/s}$. In a rough approximation the dependence follows a $1/\cos\vartheta_i$ behaviour (dashed curve).

The angular dependence of the total cross-section for diffuse scattering, i.e., of the area $\Sigma$ (3.5) associated with one admolecule is due not only to intrinsic, physical causes, but also to a trivial geometrical one. The latter apparently led to a formal contradiction between experimental [Poelsema, Palmer, De Zwart, Comsa, 1983] and theoretical work [Jónsson, Weare, Levi, 1984a; 1984b; Jónsson, 1986; Gumhalter, Liu, 1984a; 1984b; Liu, Gumhalter, 1986; Bosanac, Sunjic, 1985]. We will try to clarify the matter by means of a simple example. Let us assume that physical causes do not lead to any angular dependence, i.e., the diffusely scattering potential is isotropic. The angular dependence of $\Sigma$ is in this case entirely geometrical and has the functional form $1/\cos\vartheta_i$. We show now how this dependence is deduced from experimental and theoretical points of view.

In the theoretical approach, each admolecule is represented by a spherical potential, the spheres being located on a mathematical plane. The number of spheres seen by the incident beam, $I_0$, varies obviously as $1/\cos\vartheta_i$. The diffuse

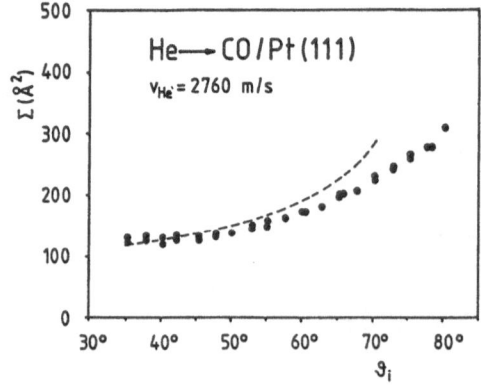

Fig. 3.7. Total cross-section or diffuse scattering of He ($v_{He} = 2760\,\mathrm{m/s}$) on CO adsorbed on Pt(111) as a function of the angle of incidence $\vartheta_i$. The dashed curve illustrates a $\cos^{-1}\vartheta_i$ behaviour

scattering of each sphere being independent of $\vartheta_i$, the relative diffuse intensity, $1 - I/I_0$, will vary as $1/\cos\vartheta_i$. Thus, in view of (3.5), the total cross section $\Sigma$ has the same functional dependence.

In the approach of the experimentalists each admolecule is represented by a hemisphere with the center located on the specularly reflecting mirror plane of the substrate. Because the integral diffuse intensity can hardly be measured, the experimentalists prefer to use the specularly reflected intensity $I$ (which for close packed metal surfaces is complementary to the integral diffuse intensity), or even the normalized value $I/I_0$. This normalized value is obviously equal to the fraction of the mirror area which is not "shadowed" by the adsorbates. Accordingly, $I/I_0$ does not depend on the number of admolecules seen by the beam, but only on the fractional area "shadowed" by the hemispheres. This area varies also as $1/\cos\vartheta_i$ so that the final result is the same. (The area $\Sigma$ is defined as that part of the mirror plane associated with the adsorbate, from which no He atom can be reflected specularly without being affected by the attractive hemisphere either on its way to or from the surface; see Fig. 3.8).

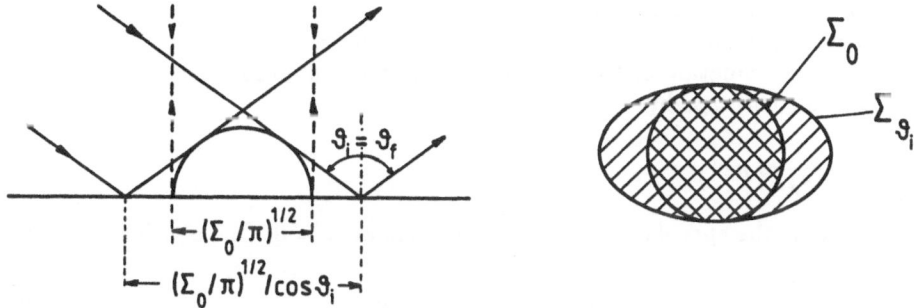

**Fig. 3.8.** Influence of the angle of incidence on the total cross-section for diffuse scattering on a hemispherical object. The in-plane cross section (*left*) and the projected surface area (*right*) are shown for perpendicular and oblique ($\vartheta_i$) incidence

The choice between the two in fact equivalent views is largely a matter of taste. As experimentalists we prefer to use for $\Sigma$ the concept of a surface area which scatters diffusely, to have the adsorbates on a real specularly reflecting mirror plane and last but not least to use as a primary concept the specular intensity, which can be directly measured, and not the diffuse intensity. For a given admolecule-substrate distance the assumption that the hemisphere centers lie on the mirror plane is exactly fulfilled only at a given He normal beam energy. Otherwise, i.e., when the centers are located above or below the mirror, the potential will be larger or smaller than a hemisphere and of course the "shadowed" area will vary with $\vartheta_i$ in a way which cannot be easily calculated, but which will certainly differ from $1/\cos\vartheta_i$. In the same case, however, the "spheres" of the theorists will also not be spheres anymore, because the presence of the reflecting mirror must then be taken into account. Thus the experimental view does not appear to be more restrictive but, at least in this latter case, more transparent.

## 3.5 Coverage Dependence of the He Reflectivity

The definition of the cross-section for diffuse scattering, $\Sigma$, given in Sect. 3.3 is purely phenomenological. It simply relates the adsorbate induced attenuation of the specular He-beam intensity, $1 - I/I_0$, with the adsorbate coverage, $\Theta$, in the limit of infinite dilution, $\Theta \ll 1$. As already mentioned, $\Sigma$ can be seen as an area on the mirror surface associated with one admolecule; in the spirit of (3.4) and (3.5) the He atoms incident on this area are scattered classically out of the specular beam. This definition is, however, not appropriate when interference effects play a non-negligible role, e.g., when the elastic scattering from admolecules has a significant contribution also in the specular direction, the phase of which differs in general from that of the scattering from the substrate.

In order to allow for interference, we redefine the cross-section for diffuse scattering again as an area on the surface associated with an admolecule, $F$, but which this time scatters diffusely the He waves; i.e., the attenuation of the specular He wave amplitude, and no more of the intensity, has to be related to the coverage. It can be demonstrated rigorously that the quantum mechanical cross-section and the classical one are simply related by $F = \frac{1}{2}\Sigma$ [Celli, 1985; Manson, 1988]. Here we will show this only in the $\Theta \ll 1$ limit. In analogy with (3.4), the attenuation of the specularly reflected amplitude $A$ is related to the admolecule coverage by

$$1 - \frac{A}{A_0} = nF = \Theta n_s F \tag{3.4'}$$

where $A_0$ is the specularly reflected amplitude at $\Theta = 0$. One obtains

$$\frac{I}{I_0} = \left(\frac{A}{A_0}\right)^2 = (1 - \Theta n_s F)^2 \simeq 1 - 2\Theta n_s F \tag{3.4''}$$

which compared to (3.4) gives $2F = \Sigma$.

Another point which has to be considered when estimating the He reflectivity, $I/I_0$, at finite coverages is the cross-section overlap. As mentioned above, for a number of adatoms and admolecules (e.g., Xe, CO, NO) the cross-section for diffuse scattering is of the order of $100\,\text{Å}^2$; even for H adatoms it is larger than $10\,\text{Å}^2$. This means that, even at coverages of a few percent, the cross-sections overlap substantially. The effect of the overlap is obvious in the adsorption curve of Fig. 3.2: the curve is not linear as expected if the cross-sections were non-overlapping (3.4 and 3.4''). The absolute value of the slope, which is proportional to the effective cross-section (3.5 and 3.5' at $\Theta \neq 0$), decreases continuously with increasing coverage; this means that the effective cross-section, $\Sigma_{\text{eff}}$, ($F_{\text{eff}}$, respectively) decreases with increasing overlap. This behaviour is well known from gas-phase scattering: the cross-section of a dimer is substantially smaller than that of two monomers [Vehmeyer et al., 1976].

There is so far no general theoretical treatment which provides a quantitative relationship between $\Sigma_{\text{eff}}$ ($F_{\text{eff}}$), at a given degree of overlap, and $\Sigma$ ($F$) defined at infinite dilution (no overlap). Such a theory should take into account effects like depolarization of individual scatterers due to their mutual interaction and multiple He scattering from neighbouring admolecules. In the absence of such a theory we have made the simplest assumption: the overlap of cross-sections is purely geometrical, while the polarizability of the individual adsorbate is not affected by the presence of the others [Poelsema, Comsa, 1985a; 1985b; Comsa, Poelsema, 1985]. In Chaps. 4, 5 and 8, the quantitative relationship between $I/I_0$ and coverage $\Theta$ based on this assumption will be compared with experimental data. The agreement achieved in a number of key cases shows that the assumption is a good approximation. Likewise, effective cross-sections calculated very recently in the sudden-approximation are within 10% of those obtained from geometrical overlap [Yinnon et al., 1988].

The degree of overlap itself, i.e. ultimately $\Sigma_{\text{eff}}$ $(F_{\text{eff}})$, depends not only on the average coverage but also on the actual lateral distribution of the admolecules (the scatterers) on the surface. It is by no means irrelevant whether a given number of adatoms is randomly distributed on the whole surface, is confined in high density islands — due to mutual attraction — or is spread apart by mutual repulsion. In the high (local) density islands the overlap is high, $\Sigma_{\text{eff}}$ $(F_{\text{eff}})$ becomes small and therefore the He reflectivity much larger than in the case of randomly distributed admolecules, corresponding to the same *average* coverage; the difference with respect to the case of repelling adsorbates is even larger. Accordingly, the coherently scattered He intensity carries valuable information on the lateral distribution of adsorbates, information which is hardly accessible with other surface probes. The procedure for extracting this information — which we call the overlap approach — will be presented and exemplified for various distributions in Chaps. 4 and 5 for homogeneous and heterogeneous systems, respectively.

A prerequisite for the use of the overlap approach is the setting up of quantitative relationships between He reflectivity, $I/I_0$, and adsorbate coverage, $\Theta$, for different kinds of adsorbate distributions. In the next two sections the detailed derivation of such relationships for a random distribution on lattice sites (lattice gas) will be presented: in Sect. 3.5.1 for perfectly diffuse scatterers and in Sect. 3.5.2 for scatterers which contribute substantially to the coherent scattering, i.e., for which terms 2 and 3 in (3.3) cannot be neglected. In Chap. 4, the $I/I_0$ vs $\Theta$ dependence will be derived for some typical distributions in the specific case of perfectly diffuse scatterers, with cross-sections much larger than the size of the unit cell of the saturated adlayer. As already mentioned, all calculations are made with the assumption that the "geometrical overlap" approach applies.

### 3.5.1 Perfectly Diffuse Scattering; Adsorbates Randomly Distributed on Lattice Sites

For perfectly diffuse scatterers, the specular He intensity originates exclusively from scattering from those parts of the surface which are not disturbed by the presence of adsorbates. The scattered amplitude is thus put equal to the non-disturbed fraction of the surface. This fraction needs to be calculated for an adsorbate coverage $\Theta$ ($\Theta = n/n_s$, where $n$ and $n_s$ denote the number of admolecules and substrate atoms per unit area, respectively) and for a disturbed area $F_A$ per adparticle. The relationship $I(\Theta)/I_0 = (A_S(\Theta)/A_0)^2$ and its derivation become particularly simple if one assumes that $F_A$ is much larger than the unit cell of the substrate; i.e. $\gg n_s^{-1}$ in the case of a primitive Bravais lattice. (For instance, for CO/Pt(111) $F_A n_s = \Sigma_A n_s/2 = 9.3$ [Poelsema, De Zwart, Comsa, 1982; 1983].) With this condition the unit cell can, in a very good approximation, be taken as the smallest area which scatters specularly. In order to ensure that a given unit cell contributes to the specular beam, the sites located on a surface of area $F_A$ centered around this cell have to be free of adsorbates. Thus $F_A n_s$ neighbouring lattice sites must be unoccupied. In the case of adsorbates randomly distributed on lattice sites, the probability of fulfilling this requirement is $(1 - \Theta)^{F_A n_s}$, since the probability for one lattice site to be unoccupied equals $(1 - \Theta)$. Thus the scattering amplitude normalized to that of the adsorbate-free ($\Theta = 0$) surface amounts to $A_S/A_0 = (1 - \Theta)^{F_A n_s}$. The relationship between the normalized He intensity (i.e. relative peak height) and the adsorbate coverage becomes

$$I/I_0 = (1 - \Theta)^{2F_A n_s} = (1 - \Theta)^{\Sigma_A n_s} \quad . \tag{3.7}$$

This expression is the "lattice gas formula".

Expression (3.7) is the surface analogue of the Lambert–Beer law in gas phase scattering: $I/I_0 = \exp(-\sigma x N)$ in which $x$ denotes the path length covered by the He atoms through a 3D-gas of density $N$ and $\sigma$ is the free particle total cross-section for diffuse scattering. The analogy would become almost an identity if the adsorbates were distributed fully randomly as in the gas phase (not on lattice sites and superposed admolecules also allowed). Equation (3.7) would then be $I/I_0 = \exp(-\Sigma \Theta n_s) = \exp(-\Sigma n)$. This expression is made identical to the gas phase one by setting the surface density $n$ equal to the 3D-density $N$ "projected" on a surface via the He flight path $x$ ($n = Nx$).

So far, it has been assumed that the adsorbates can occupy any lattice site. In many instances, however, adsorption on the nearest-neighbour sites of an already occupied site is forbidden: the closest approach of admolecules exceeds the lattice parameter. If, for instance, the occupation of nearest neighbour sites is excluded on a triangular substrate (the (111)-face of an fcc-crystal), the maximum coverage amounts to $\Theta = 1/3$. This leads to a modification of the lattice gas expression. In a good approximation the corresponding relationship $I(\Theta)/I_0$ reads: $I/I_0 = (1 - 3\Theta)^{\Sigma_A n_s/3}$, or in general

$$I/I_0 = (1 - m\Theta)^{\Sigma_A n_s/m} \quad . \tag{3.8}$$

The lattice gas formulae (3.7) and (3.8) are valid for large cross-sections ($F_A n_s \gg 1$), of perfectly diffuse scattering adsorbates, distributed randomly on step sites. The number of systems complying with these apparently restrictive conditions is actually surprisingly large (e.g. CO, NO, Xe, Kr on Pt(111), Ni(111)). As will be seen in the next chapters, the experimental data are described remarkably well by these formulae.

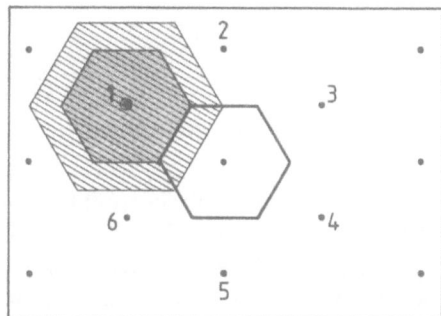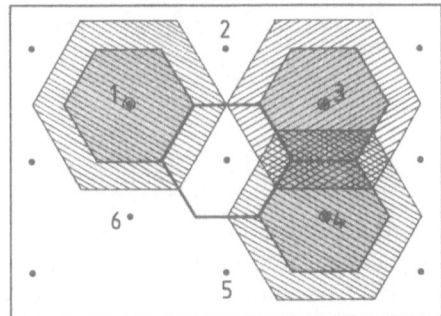

**Fig. 3.9.** Disturbance of a central hexagon (unit cell) by the presence of adsorbates on neighbouring lattice sites (adsorbates are located on site *1* (*left*) and on sites *1*, *3* and *4* (*right*) and are represented by dotted hexagons). The cross-section for diffuse scattering on an adsorbate is illustrated by the hatched hexagons

In order to be able to describe additional cases we will now drop the condition $F_A n_s \gg 1$. Let us consider a triangular lattice as substrate. For the sake of simplicity, we assume here without substantial loss of generality that both the unit cell and the scattering cross-section $F_A$ can be represented by hexagons. This is illustrated in Fig. 3.9 for $F_A n_s = 9/4$ with the adsorption sites marked by black dots. The first formula derived below applies for $1 \leq F_A n_s \leq 9/4$ and is calculated as follows.

The non-disturbed fraction of the surface at a given coverage $\Theta$ is equal to the probability averaged non-disturbed fraction of any unit cell. In order to calculate this fraction let us focus on the central unit cell in Fig. 3.9. If this cell is occupied (probability $\Theta$), the whole cell is disturbed ($F_A \geq 1/n_s$) and there is no contribution to the non-disturbed fraction; thus only situations with the central cell vacant (probability $1 - \Theta$) will be considered. The neighbouring six sites (denoted $1-6$ in Fig. 3.9) can be occupied by zero to six admolecules. For various cases the non-disturbed fraction of the central cell is different, but can be easily calculated. (For two, three and four occupied sites the disturbance depends also on the particular arrangement of the admolecules around the central cell.) The probability of each configuration (defined by the number of neighbours and their arrangement), which depends obviously on the coverage $\Theta$ but also on the type of distribution of the admolecules, has to be calculated also. The probability averaged non-disturbed fraction is then obtained by summing up the non-disturbed fraction corresponding to each configuration weighted by the probability for this configuration to occur. The result represents the normalized scattering amplitude $A_S/A_0$ from this surface. For a random distribution on lattice sites of adsorbates this amounts to

$$\frac{A_S}{A_0} = 1 - \Theta\eta^2 + \Theta^2(1 - 3\eta)(1 - \eta) - 2\Theta^3(1 - \eta)^2 \quad , \tag{3.9}$$

for $1 \leq \eta = \sqrt{F_A n_s} \leq \frac{3}{2}$. The corresponding result for $\frac{3}{2} \leq \eta \leq 2$ yields

$$\frac{A_S}{A_0} = 1 - \Theta\eta^2 + \Theta^2(7\eta^2 - 16\eta + 10) - \Theta^3(10\eta^2 - 28\eta + 20)$$
$$+ \Theta^4(4\eta^2 - 12\eta + 9) \quad . \tag{3.9'}$$

(Note that for $\eta = 2$ i.e., $F_A n_s = 4$, the result $A_S/A_0 = (1 - \Theta)^4$ is identical to that of (3.7): $I/I_0 = (A_S/A_0)^2 = (1 - \Theta)^8$).

The narrow validity range of (3.9) and (3.9') is not particularly troublesome. Firstly, because exact formulae can be easily deduced for $\eta > 2$ and $\eta < 1$. In particular, for the latter the result is trivial: $A_S/A_0 = (1 - \Theta) + \Theta(1 - \eta^2) = 1 - \Theta\eta^2$. Secondly, a more elaborate calculation for $\eta > 2$ is superfluous since its result differs only marginally from the dependency described by (3.7).

### 3.5.2 Adsorbates with Non-Negligible Reflectivity

In contrast to the CO-case, the contribution of light adsorbates, like H, to the coherent (here specular) He scattering cannot be neglected. We will quantify this as follows: if the He intensity specularly reflected from a saturated adlayer $I_1 = I(\Theta = 1)$ is larger than $0.01 \cdot I_0$, this contribution can no longer be neglected. The cross-sections for diffuse scattering measured so far being all larger than the unit cell ($\eta = \sqrt{F_A n_s} > 1$), this coherent contribution comes exclusively from the adlayer; it corresponds obviously to the third term in (3.3). The reflectivity of the saturated adlayer is $\varrho = I_1/I_0$ ($\sqrt{\varrho} = A_1/A_0$). In order to calculate the contribution of the coherent scattering from the adsorbates to the specular beam at arbitrary coverage we take advantage of an experimental observation presented in Chap. 8: vacancies in the saturated adlayer (i.e. absent adatoms) scatter the He beam diffusely very much like isolated adatoms. Accordingly, a cross-section for diffuse scattering $F_V$ can be associated to a vacancy and a normalized scattering amplitude in the specular direction $A_A/A_1$ as a function of coverage can be calculated. The calculation is analogous to that in the preceding paragraph (see also Fig. 3.10) except for the fact that the vacancy "coverage" is $(1 - \Theta)$ and that the central unit cell, whose probability averaged reflectivity is calculated, has to be occupied by an admolecule (probability $\Theta$). For a random distribution of adlayer vacancies and $1 \leq \xi = \sqrt{F_V n_s} \leq 3/2$ one obtains

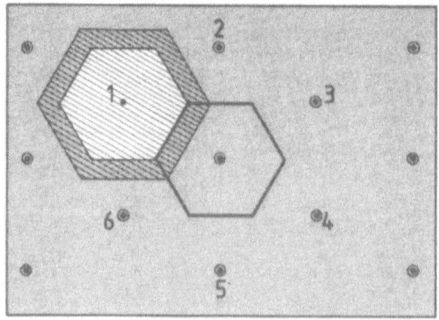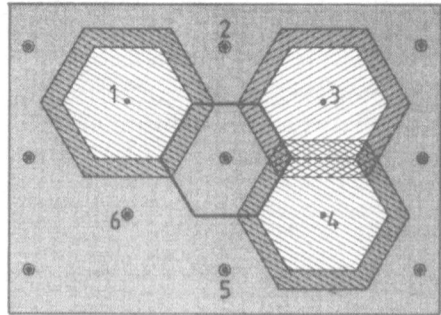

Fig. 3.10. Analogon of Fig. 3.9 for diffuse scattering from vacancies

$$\frac{A_A}{A_1} = \frac{A_A}{\sqrt{\varrho}\,A_0} = \Theta(2-\xi)^2 + \Theta^2(\xi-1)(5-3\xi) + 2\Theta^3(1-\xi)^2 \quad . \tag{3.10}$$

The corresponding result for $\xi$ ranging from $3/2$ to $2$ yields

$$\frac{A_A}{A_1} = \frac{A_A}{\sqrt{\varrho}\,A_0} = \Theta(\xi-2)^2 + \Theta^2(\xi-2)^2 - 2\Theta^3(\xi-2)(3\xi-4)$$
$$+ \Theta^4(2\xi-3)^2 \quad . \tag{3.10'}$$

These formulae are equivalent to expressions (3.9) and (3.9') for adatoms. The amplitude $A_A$ is normalized with respect to $A_1$, the amplitude specularly reflected from the saturated adlayer, which is related to $A_0$ by $A_1 = \sqrt{\varrho}\,A_0$. Figure 3.11 illustrates the combined situation: the white and the dotted areas are reflecting specularly of course with different reflectivities (1 for the substrate and $\varrho$ for the adlayer, respectively); the hatched areas scatter diffusely, due to both vacancies and adatoms.

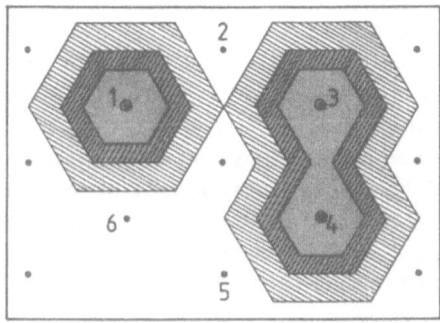

Fig. 3.11. Illustration of the scattering contributions into the specular direction from a substrate partly covered with adsorbates. The dashed areas scatter diffusely. The free and adsorbate-covered (*dotted*) areas contribute with different reflectivities to the specular He beam

The total scattering amplitude is obtained by combining expressions (3.9) and (3.10):

$$A = A_S + A_A\,e^{i\varphi} \quad , \tag{3.11}$$

where $\varphi$ denotes the phase difference between scattering contributions from clean and adsorbate-covered parts of the surface. The relative peak height is then obviously given by expressions (3.9) and (3.10) inserted in (3.3'). This formula permits a detailed quantitative description of complicated adsorption systems, as will be shown in Chap. 8 for H/Pt(111).

The initial slope of the adsorption curve is given by a combination of expressions (3.9, 3.10, 3.3') or, equivalently, (3.9', 3.10', 3.3'):

$$\left.\frac{dI/I_0}{d\Theta}\right|_{\Theta=0} = -2[\eta^2 - (2-\xi)^2\sqrt{\varrho}\,\cos\varphi] \quad . \tag{3.12}$$

Thus, when the adsorbates do reflect specularly, the initial slope of the adsorption curve is determined not only by the total cross-section for diffuse scattering, but also by coherent scattering from the adsorbates. Consequently, a simple direct determination of the cross-sections for diffuse scattering from the initial slope of the adsorption curves, as for perfectly diffuse scatterers, is no longer possible but requires a detailed evaluation of the experimental data in the entire coverage range (see also Chap. 8).

Note from (3.12) that, for $\xi = 2$, $d(I/I_0)/d\Theta = -2\eta^2 = -2F_A n_s = -\Sigma_A n_s$. In other words $\Sigma_A$ can be deduced directly from the initial slope of the adsorption curve in the particular case $\xi = 2$. This applies more generally for $\xi > 2$. Indeed, the existence of an isolated adparticle requires by definition that all neighbouring sites are vacant. Since, for $\xi \geq 2$, the cross-section of one single neighbouring vacancy covers the complete central unit cell, an isolated reflecting adparticle will act as a perfect diffuse scatterer: no scattering contributions into the specular direction have to be taken into account.

### 3.5.3 First Layer Vacancies

A particular case of diffusely scattering objects with a high reflectivity is represented by vacancies in the first substrate layer. As will be discussed in Chap. 4 these vacancies have large cross-sections for diffuse scattering; they are of the same order as those for adatoms. On the other hand, the reflectivity of a layer with vacancy density $= 1$, e.g. just after complete removal of the first layer by sputtering (see Chap. 6), is obviously identical to that of the ideal first layer: $I/I_0 = \varrho = 1$. For randomly distributed vacancies one may now calculate $A/A_0$ according to the procedure outlined above. For the sake of simplicity we take the realistic case of an equal cross-section of $\Sigma = 18n_s^{-1}$ for diffuse scattering from both a vacancy and a substrate adatom (see Sect. 4.4). The scattering amplitude from the undisturbed fraction of the first layer then becomes

$$\frac{A}{A_0} = (1 - \Theta)^9 \quad . \tag{3.13}$$

The scattering amplitude from the uncovered second layer, i.e. from the vacancies, yields

$$\frac{A}{A_0} = \Theta^9 e^{i\varphi} \quad , \tag{3.14}$$

$\varphi$ being the phase difference between scattering contributions from the first and second layer. Up to a vacancy concentration $\Theta = 0.374$ the contribution of (3.14) to the total scattering amplitude is at least two orders of magnitude smaller than that of (3.13). Consequently, a description in terms of (3.13) alone is fully appropriate up to this concentration. For deviations from a random distribution the situation is different, as is amply discussed in Chap. 6. This can be easily checked experimentally by variation of the phase difference $\varphi$ (change of angle of incidence or energy).

# 4. Diffuse Scattering as a Tool to Probe the Lateral Distribution of Adatoms and Vacancies (Homogeneous Systems)

## 4.1 The Overlap Approach

As shown in the preceding chapter, the large size of the cross-section for diffuse scattering (diameter $\sim 10\,\text{Å}$) leads even at low coverages to a substantial overlap of the cross-sections of the individual scatterers. It was further argued that at a given coverage the degree of overlap depends on the actual distribution of the scatterers along the surface, and thus that this distribution determines the shape of the dependency of the relative height $I/I_0$ of the specular He peak on surface coverage. This opens the way for a novel application of TEAS as a surface probe: the determination of the lateral distribution of the adsorbates, and thus of their mutual interaction, from the measurement of $I/I_0$ vs $\Theta$. This overlap approach will be discussed and exemplified in this section, in the simple but widespread case of purely diffuse scatterers (negligible specular reflection from the full adlayer).

Three types of distributions resulting from three limiting cases of interaction between admolecules are considered here: (1) no interaction, (2) maximum repulsion and (3) maximum attraction. For perfectly diffuse scatterers each case results in a simple $I/I_0$ vs $\Theta$ dependency.

**1) No Interaction (l.g.):** The molecules are distributed randomly on lattice sites (lattice gas). This case has been discussed in detail in Sect. 3.5.1. For large cross-sections one obtains (3.7)

$$\left.\frac{I}{I_0}\right|_{\text{l.g.}} = (1 - \Theta)^{\Sigma n_s} \quad . \tag{4.1}$$

A lattice gas distribution is obtained also for interacting admolecules if they stay immobile at the impact site when adsorbed from the gas phase. This may happen at very low surface temperatures if the sticking molecules lose their kinetic energy instantaneously.

Formula (3.8), with $m = 3$, has been deduced for a lattice gas, when the occupation of nearest neighbour sites has been excluded. This case can be viewed as strong repulsion between neighbouring admolecules and no interaction beyond nearest neighbour sites.

**2) Maximum Repulsion (m.r.):** The repulsion is large enough to exclude any overlap of the scattering cross-sections. The highest attainable coverage is obviously $\Theta_{\max} = (n_s F)^{-1}$. The overlap being excluded, the $I(\Theta)/I_0$ dependency is identical to that for infinite dilution (3.4″)

$$\frac{I}{I_0}\bigg|_{\text{m.r.}} = (1 - n_s F \Theta)^2 \quad . \tag{4.2}$$

**3) Maximum Attraction (m.a.):** On a defect-free surface, the adsorbates build a single island with density $1/U$ ($U$ is the size of the corresponding unit cell in the close-packed adsorbate island). Obviously, adsorbate mobility is a prerequisite for this behaviour. In the case of a unique (or few) island(s), its size becomes even at very low coverages $\Theta < 0.001$ much larger than the transfer width and thus the problem may be treated classically. The $I(\Theta)/I_0$ dependency is then

$$\frac{I}{I_0}\bigg|_{\text{m.a.}} = 1 - n_s U \Theta \quad . \tag{4.3}$$

At higher defect concentration (where the defects act as island nucleation centers), there are many but small islands even at finite coverage. In the limit of island sizes, which are small compared to the transfer width, one should use

$$\frac{I}{I_0}\bigg|_{\text{m.a.}} = (1 - n_s U \Theta)^2 \quad , \tag{4.4}$$

instead of (4.3). In the case of small islands, edge effects complicate the situation additionally (see also Chap. 6).

Figure 4.1 shows the relationship between the relative peak height $I/I_0$ and the adsorbate coverage derived above for different lateral distributions. The following parameters are used: $n_s^{-1} = 6.67\,\text{Å}^2$, $U = 20\,\text{Å}^2$ and $\Sigma = 120\,\text{Å}^2$. A plot of measured $I(\Theta)/I_0$ data yields directly the nature of the adsorbate-adsorbate interaction: data in ranges (I) and (II) indicate the presence of repulsive or attractive interactions, respectively. Quantitative information on the lateral distributions of the diffuse scatterers (adsorbates and/or defects) is inferred by fitting

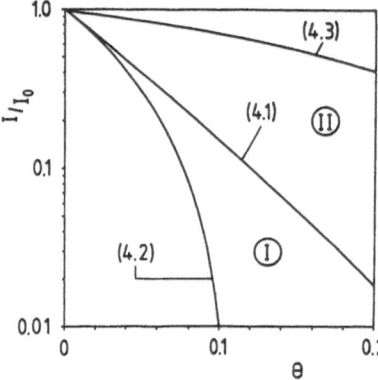

Fig. 4.1. Calculated relative He specular peak heights as a function of coverage for various lateral distributions of diffuse scatterers: lattice gas (4.1), maximum repulsion (4.2) and maximum attraction (4.3). The calculations are performed for $n_s^{-1} = 6.67\,\text{Å}^2$, $A = 20\,\text{Å}^2$ and $\Sigma = 120\,\text{Å}^2$ (see text)

$I(\Theta)/I_0$-curves obtained from appropriate model calculations to the experimental data. Examples for CO/Pt(111), vacancies on Pt(111) and Xe/Pt(111) are shown below.


## 4.2 The Applicability of the Overlap Approach


In the previous section the overlap approach has been introduced for perfect diffuse scatterers. As already mentioned, this situation applies in good approximation for a great number of adsorbates such as CO/Pt(111). The adsorption curve in Fig. 3.2 shows that the intensity of the specular He beam from the saturated CO adlayer is weaker by about 3 orders of magnitude than that from the clean substrate. This applies also, e.g., for NO, $O_2$ and Xe on Pt(111) at temperatures above 80 K. The large probability ($> 99\%$) for diffuse scattering from saturated adlayers is attributed to the softness (i.e. low Debye temperature) of the adlayer. In the case of well-ordered adlayers, the remaining coherent scattering leads to well-defined diffraction patterns. These can be investigated only using TEAS instruments with a large dynamical range ($5-6$ orders of magnitude), see e.g. [David, Kern, Zeppenfeld, Comsa, 1986].

However, for the investigation of the lateral distribution of adsorbates, in particular, and of the disordered layers, in general, this is not necessary. Since the information is obtained from the attenuation of the specular He-beam originating from the uncovered areas of the substrate, a TEAS apparatus with a dynamical range of $2-3$ orders of magnitude, as described in Chap. 2, is fully adequate. In this case, of course, even the coherent scattering from well-ordered adlayers of CO, NO, $O_2$ and Xe merges into the background. Consequently, these adsorbates can be considered as perfectly diffuse scatterers and the overlap approach applies in the entire coverage range.

If instead of the admolecules mentioned above, the scatterers are metal-adatoms or vacancies induced by ion sputtering the situation may be radically different. Let us assume that metal-adatoms are adsorbed on a substrate made up of the same metal (e.g. Pt atoms on a Pt(111) surface). If the surface temperature is not too low the metal-adatoms are mobile and aggregate in islands (2D-clusters) of the same structure as the substrate. At full coverage ($\Theta = 1$) the adlayer is obviously identical with the original "non-covered" substrate ($\Theta = 0$) and its reflectivity also. This means that the $I/I_0$ does not merge into the background at high coverages but has to recover at full coverage its original value $I/I_0 = 1$. In between, the intensity passes through a minimum; the shape of the $I(\Theta)/I_0$ curve depends mainly on the interference between the He waves scattered from the uncovered substrate and from the highly reflecting islands. This behaviour will be discussed in detail in Chap. 6 together with the practically identical behaviour of vacancies at higher temperatures. Indeed, phenomenologically, metal-adatoms and vacancies behave alike except for the fact that the vacancies may stay immobile up to higher, experimentally more comfortable temperatures [e.g. $\sim 180$ K on Pt(111)]. The overlap approach for purely diffuse scatterers as introduced in the preceding section can be used for metal-adatoms and vacancies when the interference effects can be neglected; i.e. as long as the islands are smaller than the cross-section of an individual metal-adatom or vacancy. This is only the case at low temperatures (immobility) and not too high coverages. An example will be given in Sect. 4.4 below for "immobile" vacancies.

Thermal He scattering from light adatoms like H is in fact more similar to the scattering from metal-adatoms than from the heavier adatoms or admolecules. Indeed, due to the strong binding on the substrate and the small mass, the stretch-frequency of the H/substrate bond is much too large to be excited by thermal He scattering. Even the wagging modes are hardly excited, both because their associated energies are rather high and because the excitation of parallel vibrations by thermal He impact appears to be improbable. As a consequence the Debye-temperature of the saturated H monolayer is only slightly smaller than that of the clean substrate (e.g. 190 K and 230 K for H/Pt(111) and Pt(111), respectively) [Engel, Kuipers, 1979; Lee, Cowin, Wharton, 1983; Poelsema et al., 1986]. The high reflectivity of the H adlayer leads to interference effects, which complicate the $I/I_0$ vs $\Theta$ curve. However, as will be shown in Chap. 7, in this case also the $I(\Theta)/I_0$ dependency can be well modelled (by using the formulation discussed in Sect. 3.5.2) and used to obtain information on the H adlayer.

The examples in the next sections of this chapter are confined to the simplest situation of large, purely diffuse scatterers. We will consider both repulsive and attractive interactions between scatterers.

*Final Remark:* the decision whether the overlap approach for purely diffuse scatterers may be used can be obtained directly from the experiment. The simplest but most restrictive criterion is the practical disappearance of coherent scattering at full coverage. A more specific criterion applicable at all coverages is to look for the presence or absence of interference effects by varying the scattering conditions (He incidence angle and/or energy) — see, e.g., Chaps. 6 and 8.

## 4.3   Repulsive Interactions: CO/Pt(111)

It has been shown that CO does not form islands at any temperature on Pt(111) (see [Poelsema, Verheij, Comsa, 1982; 1983b] and e.g., Sect. 5.1). Thus its total cross-section for diffuse scattering $\Sigma$ is obtained from the initial slope of the adsorption curve $I/I_0$ vs CO exposure and the initial sticking coefficient. $\Sigma_{CO}$ amounts to 123 Å$^2$ for a 63 meV He beam incident at $\vartheta_i = 40°$ [Poelsema, De Zwart, Comsa, 1982; 1983].

For randomly distributed CO/Pt(111), the lattice gas behaviour according to (4.1) is expected. Such behaviour is illustrated by the solid curve in Fig. 4.2 which has an initial slope corresponding to $\Sigma_{CO} = 123$ Å$^2$. However, the experimental data points, denoted by crosses, deviate even at low coverages from the solid curve. The experimental relative specular peak height $I/I_0$ decays faster than would be predicted by the lattice gas model. Keeping in mind the discussion

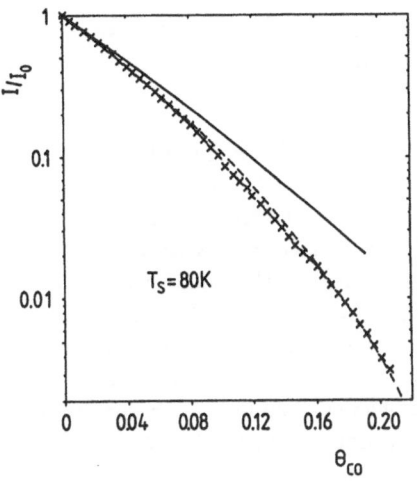

**Fig. 4.2.** Relative He specular peak height as a function of CO coverage. The experimental data (×) were taken with a 63 meV He beam incident at $\vartheta_i = 40°$ on Pt(111). The curves are calculated from (4.1) for a lattice gas distribution (*solid curve*) and from (4.5) for a lattice gas with exclusion of occupation of nearest neighbour sites (*dashed curve*)

of Fig. 4.1, this fact allows one to classify straightforwardly the nature of the CO-CO interaction as being repulsive.

More detailed information on the lateral distribution of CO is obtained by modelling a distribution based on CO-CO repulsion and deriving the corresponding $I/I_0$ vs $\Theta$ dependence; this is then compared to the experimental data. Electron energy loss spectroscopy measurements have shown that for $\Theta_{CO} < 0.17$, all CO molecules are adsorbed on on-top sites [Steininger, Lehwald, Ibach, 1982; Steininger, 1982]. Thus the simplest "repulsive" model assumes that the CO molecules occupy randomly the on-top sites except for the nearest-neighbour on-top sites of an already adsorbed molecule. Thus, as shown in Sect. 3.5.1, for large scattering cross-sections the relative peak height $I/I_0$ vs $\Theta$ is given by

$$I/I_0 = (1 - 3\Theta)^{n_s \Sigma/3} \quad , \tag{4.5}$$

i.e. (3.8) with $m = 3$. The plot of (4.5) shown in Fig. 4.2 (dashed curve) agrees surprisingly well with the data in view of the simplicity of the model. Above $\Theta_{CO} = 0.17$ the dashed curve given by (4.5) might actually deviate downwards with respect to the data. Indeed, again electron energy loss measurements indicate that above $\Theta_{CO} = 0.17$, the CO molecules start to occupy bridge sites also [Steininger, Lehwald, Ibach, 1982], i.e. they may come nearer to each other. As a consequence the overlap may become relatively larger and the $I/I_0$ decrease with $\Theta_{CO}$ somewhat slower. Both the uncertainty of the data (e.g. slight variation of CO sticking coefficient with coverage) and the crudity of the model do not allow so far a confirmation of this tendency.

## 4.4 Attractive Interaction Between Thermally Immobile Vacancies on Pt(111) at $T_s = 80$ K

Via the overlap approach, TEAS also provides useful insight into the lateral distribution of ion bombardment induced vacancies. Such information is not at all or hardly accessible to conventional techniques. In this section, this TEAS application is exemplified for vacancies on Pt(111) at $T_s = 80$ K [Poelsema et al., 1985]. This temperature is sufficiently low for keeping the vacancies thermally immobile. This has been inferred from the fact that below 180 K the relative He specular peak height $I/I_0$ does not change after switching off the damaging ion beam, but it does change if the surface temperature is 180 K or higher.

Note, that the monitoring of the state of the surface by TEAS can be done continuously, i.e., in particular also during ion damaging; the thermal He beam and the damaging ion beam do not interfere in any way.

Figure 4.3 shows typical damaging curves, i.e. the relative peak height $I/I_0$ as a function of the vacancy density $\Theta_d$. The vacancy density is given by the product of the ion fluence times the sputtering rate. The fluence is monitored directly during the measurements. The sputtering rate, determined according to the method described in Chap. 6 [Poelsema, Verheij, Comsa, 1984a], is assumed to be temperature-independent [Roosendaal, 1981]. For 400 eV $^3$He$^+$ ions (results denoted with $\times$ in Fig. 4.3), the sputtering yield amounts to $\sim 0.02$ atoms/ion. At such low rates, the probability to sputter more than one Pt atom upon impact of one $^3$He$^+$ ion is negligibly small ($< 3 \times 10^{-4}$). Consequently, the vacancies are created one at a time, i.e. independent of each other. At low densities, the vacancies are well separated. Therefore, the condition for the determination of the total cross-section for diffuse scattering (isolated scatterers) is fulfilled. The initial slope of the damaging curve in Fig. 4.3 yields a total cross-section $\Sigma = 2F_V = 150$ Å$^2$ for He diffuse scattering from a monovacancy ($E_{He} = 16$ meV, $\vartheta_i = 40°$). This value is similar to those for adsorbates, suggesting that in this case too, long-range dispersion forces are the main cause

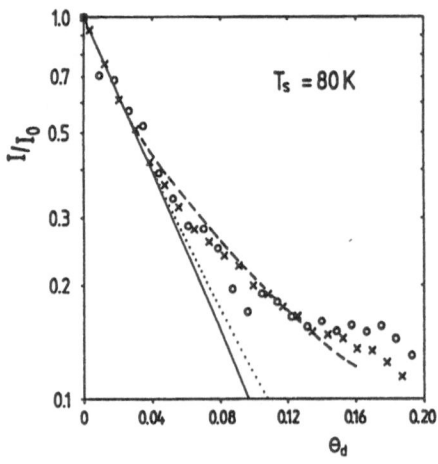

**Fig. 4.3.** Damaging curves: the relative He specular peak height ($E_{He} = 16$ meV, $\vartheta_i = 40°$) as a function of the sputtered fraction of the Pt(111) outermost surface-layer at $T_s = 80$ K. The data were taken with 600 eV Ar$^+$ (o) and 400 eV $^3$He$^+$ ($\times$) primary ions. The solid curve corresponds to a lattice gas distribution (4.1). The dotted and dashed curves are obtained by allowing a newly created vacancy to make one and two jumps, respectively, in the direction of an already existing vacancy

[Poelsema et al., 1985] (see Chap. 3). Calculations performed by Zaremba [1985] and Yinnon et al. [1988] strongly support this picture. This fact enables the application of the overlap approach described above. The solid curve shows the lattice gas dependence (4.1) with $\Sigma = 150\,\text{Å}^2$. The upward deviation of the data from this behaviour indicates immediately (compare to Fig. 4.1 and the corresponding discussion) that the interaction between (mono-) vacancies is attractive. This attraction provides obviously the basis for the well known damage annealing process occurring at higher temperatures.

The actual lateral distribution of vacancies can also be inferred — as for CO in the preceding paragraph — from the analysis of the detailed shape of the damaging curve. For densities up to $\Theta_d = 0.04$ the damage data lie on the lattice gas curve. Thus the vacancies are randomly distributed, obviously, on lattice sites. This had to be expected in view of the random impact of the $^3\text{He}^+$ ions and of the thermal immobility of the created vacancies at $T_s = 80\,\text{K}$. The upward deviation at higher densities due to the enhanced degree of overlap indicates that the mean intervacancy distance decreases with respect to the random one due to the attractive interactions, in spite of the established thermal immobility. This can be rationalized by means of a simple model: immediately upon creation, the monovacancies are allowed to make a maximum number of $n$ lattice jumps, into the direction of an already existing vacancy. The jumps become possible due to the excess energy transferred to the lattice by the ion impact. The result of model calculations for $n = 1$ and $n = 2$ is shown in Fig. 4.3 by the dotted and the dashed curves, respectively. The dashed curve describes fairly well the experimental data, indicating that the nascent vacancies jump at most twice into the direction of an already existing vacancy (-cluster). The corresponding lateral distribution for a vacancy concentration $\Theta_d = 0.10$ is schematically shown in Fig. 4.4. Note that the restricted mobility has no thermal component but is exclusively enabled by the excess energy transferred to the lattice during vacancy creation. This is proven by the fact that the shape of the damaging curve is temperature independent up to $T_s \approx 180\,\text{K}$.

In the case of higher sputtering yields, the probability that two or more Pt atoms are sputtered upon the impact of a single ion is no longer negligible. The question of whether or not nearest-neighbour Pt atoms are ejected in one sputtering event, i.e. whether the vacancies created by one incident ion are clustered

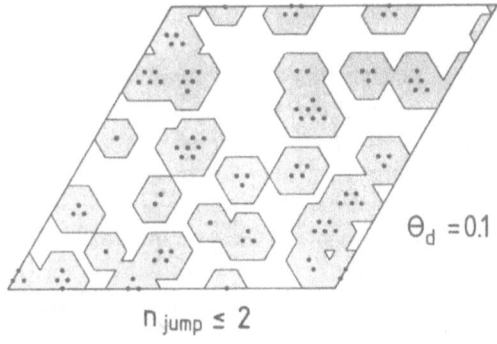

$\Theta_d = 0.1$

$n_{jump} \leq 2$

**Fig. 4.4.** Schematic view of the vacancy distribution, at a vacancy concentration $\Theta_d = 0.1$, for a $n \leq 2$ jump model (see text)

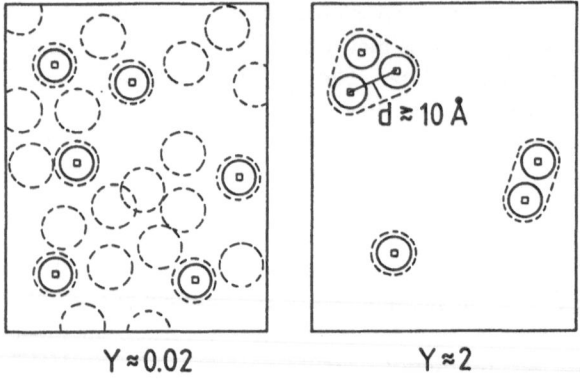

**Fig. 4.5.** Schematic illustration of the lateral distribution of vacancies created with sputtering yields of $Y \approx 0.02$ (*left*) and $Y \approx 2$ atoms/ions (*right*). The small squares represent monovacancies, the solid circles their associated scattering cross-sections and the dashed enclosures sketch single impact events

or not, is of particular interest. In order to shed light on this question, measurements have been performed with 600 eV $Ar^+$ ions. These particles have a sputtering yield of 2.5–3 Pt atoms/ion. This implies that the probability to create more than one vacancy as a result of a single ion impact event is larger than 50%. The corresponding damaging curve (○) is also plotted in Fig. 4.3 and can be compared with the one for 400 eV $^3He^+$ ions (×). The curves are identical within experimental uncertainties. The two induced vacancy distributions are indistinguishable to thermal energy He atoms. This means that at a given average vacancy density the creation of several vacancies in one ion impact event does not result in a larger *local* vacancy density, at least at the scale corresponding to scattering cross-sections of $F_V = 75\,\text{Å}^2$. Therefore, one may infer that the average distance between vacancies created in a single ion impact event is larger than the characteristic length of 10 Å ($\approx \sqrt{4F_V/\pi}$); i.e. the Pt atoms ejected in a single sputtering event are not nearest neighbours. A schematic representation of both distributions is shown in Fig. 4.5.

## 4.5 Attractive Interaction Between Mobile Adsorbates: Xe/Pt(111)

The Xe/Pt(111) adsorption curve, i.e. the relative He specular peak height $I/I_0$ versus Xe exposure shown in Fig. 4.6 has a very characteristic shape: the initial sharp decrease of $I/I_0$ is interrupted by an evident kink, the decrease becoming about six times slower and, over a large exposure range, *linear* (note that, in contrast to the preceding $I/I_0$ vs "exposure" plots, the ordinate in Fig. 4.6 is linear). The obvious upward bending of the adsorption curve shows that, like the vacancies but unlike the CO molecules, the Xe atoms attract each other. In view of the discussion in Sect. 4.1 (maximal attraction), the substantial linear

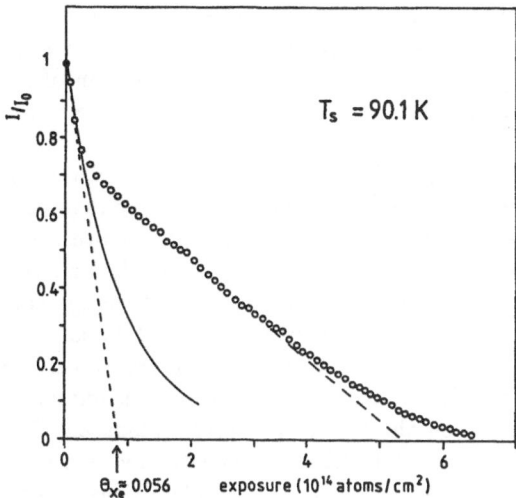

Fig. 4.6. Relative height of the He specular peak as a function of exposure measured during Xe adsorption on Pt(111) at $T_s$ = 90 K. ($E_{He}$ = 63 meV and $\vartheta_i$ = 40°). The solid curve corresponds to the lattice gas expression (4.1) ($\Sigma$ = 120 Å$^2$) and the dashed one indicates the initial slope of the curve

segment of the adsorption curve also indicates (assuming that the Xe sticking probability is constant), that Xe forms large islands. This shows that unlike the vacancies which are immobile up to 180 K, the Xe atoms are very mobile in the whole range investigated here: 81.6−91.6 K. (It has been confirmed recently that already above 40 K Xe is fully mobile on Pt(111) [Kern, David, Palmer, Comsa, 1986].) The existence of large Xe islands has been demonstrated directly by diffraction experiments: a sharp $(\sqrt{3} \times \sqrt{3})R30°$ LEED pattern has been observed at coverages corresponding to the linear segment [Poelsema, Verheij, Comsa, 1983a; 1985b]. More recently, high resolution He-diffraction measurements have shown that the $(\sqrt{3} \times \sqrt{3})R30°$ structure sets in just beyond the kink and that the Xe islands may reach sizes of the order of 1000 Å [Kern, David, Palmer, Comsa, 1986]. As shown by (4.3) the slope of the straight line in the adsorption curve is determined by the area $U$ of the unit cell of the adlayer. From the data in Fig. 4.6 we deduce

$$U = \frac{20 \pm 2}{s_{Xe}} \text{ Å}^2$$

where $s_{Xe}$ is the Xe sticking probability. As the unit cell of the $(\sqrt{3} \times \sqrt{3})$ structure on Pt(111) is $U = 20$ Å$^2$, the sticking probability turns out to be unity within the experimental error.

In analogy with the 3D-condensation, we have to assume that at a density (coverage) lower than that corresponding to the vapour pressure at the given temperature, the Xe atoms behave like a gas (in 2D, like a lattice gas). Thus we can draw through the experimental points on the left of the kink a lattice gas curve (solid curve) and can deduce from its initial slope the cross-section for diffuse scattering of Xe: $\Sigma$ = 120 Å$^2$ for 63 meV He incident at $\vartheta_i$ = 40°.

The important point which has to be emphasized here is that TEAS offers the unique opportunity to visualize directly the 2D-condensation by a substantial change in slope of the adsorption curve (here by a factor of $\sim 6$). The coverage at which the kink occurs, $\Theta_g$, is the critical density at which at the given temperature the condensation takes place. In view of the large cross-section of Xe this phase transition can be observed already in a temperature range, were the coverage $\Theta_g$ is as low as a few tenths of a percent.

Thus TEAS is well suited to investigate in detail the 2D-condensation of adsorbates. Let us illustrate this here for Xe on Pt(111). The low-coverage part of Xe adsorption curves has been measured at several surface temperatures. The result is plotted in Fig. 4.7. All curves have a common initial part (dashed curve), representing the Xe lattice gas. The $\Theta_g$ values obtained from the coverage corresponding to each kink are sufficiently small to ensure that the 2D-Xe gas can be treated as an ideal gas. Thus the 2D-Xe vapour spreading pressure can be obtained from $p_{2D} = n_s \Theta_g k T_s$. During the measurement of the adsorption curve there is obviously no equilibrium between the 3D-Xe gas phase (constant 3D-Xe pressure) and the 2D-Xe adsorbed phase (the 2D-Xe concentration increases with time). However, the shape of the adsorption curve is completely insensitive with respect to large variations of the 3D-Xe pressure. This means that the two adsorbed Xe phases (2D-gas and 2D-solid) are in dynamical equilibrium. Consequently, the data in Fig. 4.7 can supply meaningful thermodynamic values, as for instance the 2D heat of vaporization. For this purpose, the logarithm of the 2D-Xe vapour pressure is plotted versus the reciprocal surface temperature in Fig. 4.8. From the slope of the resulting straight line [Poelsema, Verheij, Comsa, 1983a] the 2D heat of vaporization is determined to be $\lambda_{2D} = 47$ meV. This is the first direct determination of the 2D heat of vaporization of adsorbed noble gas layers and the value is in agreement with the theoretical result: $\lambda_{2D} = 43$ meV [Bruch, 1983].

Note that the concentration $\Theta_g$, associated with the 2D vapour pressure for Xe, becomes extremely small at low temperatures, see also [Phillips, Bruch,

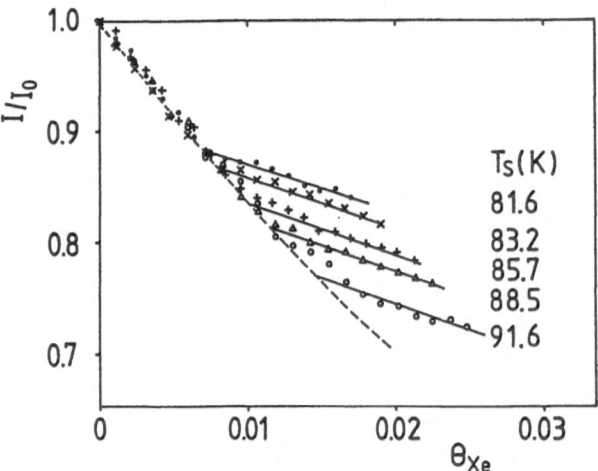

Fig. 4.7. Initial part of Xe-adsorption curves on Pt(111) at various substrate temperatures. The dashed curve exhibits the lattice-gas behaviour (4.1)

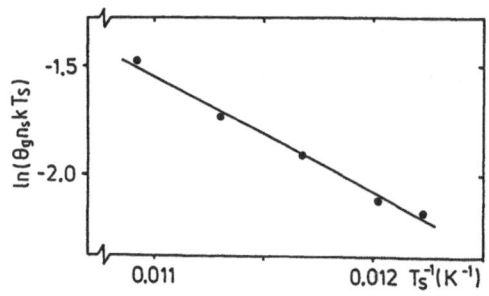

**Fig. 4.8.** Two dimensional Xe vapour pressure as a function of the reciprocal Pt(111) surface temperature

Murphy, 1981; Barker, Henderson, Abraham, 1981]; for instance $\Theta_g = 0.001$ at $T_s = 62\,\text{K}$. Such low coverages cannot be measured — not even with TEAS — and the change of slope, indicating the onset of 2D-condensation, would coincide with the origin of the adsorption curve. Thus the initial slope of the adsorption curve would be determined by $U$ ($= 20\,\text{Å}^2$), rather than by $\Sigma_{Xe}(= 120\,\text{Å}^2)$. Consequently, the cross-section for diffuse scattering from a Xe atom can only be deduced in a narrow temperature interval: it is limited on the low $T_s$ side by a non negligible $\Theta_g$, required for determining the initial slope of the adsorption curve and on the high $T_s$ side by desorption from the 2D-Xe gas phase. The low $T_s$ limit may be avoided by taking the adsorption data at sufficiently low temperatures so that the adsorbates are immobile. *Provided the atoms remain randomly distributed on lattice sites upon adsorption*, i.e. just as they arrive from the gas phase, the initial slope would again be associated with diffuse scattering from isolated Xe atoms.

# 5. Diffuse Scattering as a Probe of the Lateral Distribution in Heterogeneous Systems

The previous chapter has introduced the procedure to extract information on the lateral distribution of diffuse scatterers in homogeneous systems by means of the overlap approach. This application of TEAS is by no means restricted to homogeneous systems. Diffuse scatterers of various natures, such as defects and adsorbates or adsorbates of a different kind are often simultaneously present at the surface. The extension of the overlap approach to heterogeneous systems gives the opportunity to extract information, both on the interactions between adsorbates and defects and on those between the different components in coadsorption systems. Examples are presented and discussed in the subsequent sections.

## 5.1 Adsorption on Surfaces with Defects

Atomic steps represent probably the most common defect species on crystalline surfaces; they are certainly always present, even on surfaces which have been prepared using the latest state of the art techniques [Linke, Poelsema, 1985]. The discussion of the overlap between the diffuse scattering cross-sections of adsorbates and steps is appropriate. In general, adsorbates are bound more strongly to defect sites than to sites on ideal parts of the surface. In case the adsorbates are mobile on the defect free part of the surface, they will migrate and eventually stick at some defect; as the dwell time at defect sites is longer, these are occupied preferentially. As shown in the preceding chapter, both adsorbates [Poelsema, De Zwart, Comsa, 1982; 1983; Poelsema, Verheij, Comsa, 1983a] and defects [Poelsema et al., 1985] possess large cross-sections for diffuse scattering of thermal helium atoms; specifically, the scattering cross-section of a step row can be viewed as a band about 12 Å broad extending along the step row [Verheij, Poelsema, Comsa, 1985] (see Chap. 6 for details). Consequently, cross-sections of adsorbates bound at step sites overlap substantially with those of the steps. The bottom line is that an admolecule at a step site has in general a much smaller effective cross-section for diffuse scattering than the nominal one.

Two significant aspects concerning the cross-section overlap between adsorbates and the steps at which they are bound will be discussed here: one, with a partially negative touch — a warning concerning the experimental determination of adsorbate cross-sections, connected with a way to titrate the defect density — and the other, certainly positive — a unique method for measuring the migration of admolecules on monocrystals at low coverages.

First the warning. As noted earlier the definition of the cross-section for diffuse scattering, (3.4) as well as (3.5), is meaningful only if the scatterers are sufficiently isolated so that no overlapping of cross-sections occurs. Only as long as this condition is fulfilled, will each newly arriving adsorbate contribute with its full nominal cross-section to the further reduction of the He reflectivity. However, this condition is certainly not fulfilled if defects are present on the surface and the adsorbates migrate to the defect sites immediately upon adsorption; their cross-sections overlap from the very beginning. If so, already from the start of the exposure each newly arriving adsorbate will contribute to the reduction of the reflectivity only with that fraction of its cross-section which has not overlapped with the cross-section of the step on which the adsorbate is bound. This fraction, which is the effective cross-section for diffuse scattering of the adsorbate under the given experimental conditions (non-negligible amount of defects present and mobile adsorbate), is in general substantially smaller than the nominal cross-section of the adsorbate. As a consequence, the initial slope of the adsorption curve $d(I/I_0)/d\Theta|_{\Theta=0}$ on a surface with defects is substantially smaller than that on a defect free surface. This is illustrated in Fig. 5.1: both curves, CO/Pt(111) are measured at $T_s = 293\,\mathrm{K}$, however, $(a)$ is on a defect free surface $(\times)$ while $(b)$ on one with defects $(\square)$. The warning is now obvious. The cross-section determined from the initial slope of curve $(b)$ would be $\Sigma = 42\,\text{Å}^2$ instead of the correct nominal value of $123\,\text{Å}^2$ obtained from curve $(a)$.

Before discussing ways to avoid this type of error let us comment for a moment on the full shape of curve $(b)$ in Fig. 5.1. After a certain exposure the data points leave the initial slope reaching eventually a slope parallel to that of curve $(a)$. This happens when the defects become saturated with CO. The CO molecules adsorbed after defect saturation stay on the defect free surface and induce a behaviour similar to that of curve $(a)$. This observation points to another interesting application of TEAS, which is in some cases very helpful: the *titration* of defects. Indeed, the coverage corresponding to the intersection of the two slopes can be taken as the amount of CO necessary to saturate the defects. Making a reasonable assumption for the number of CO molecules nec-

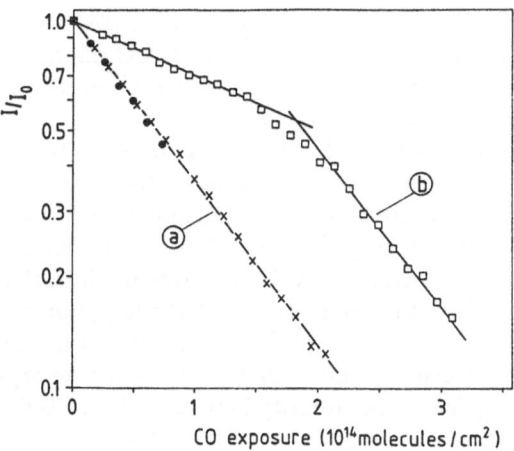

Fig. 5.1. Relative He specular peak height as a function of exposure for CO/Pt(111). The data, obtained with a 63 meV He beam incident at $\vartheta_i = 40°$, were taken with a "perfect" (defect concentration $\approx 10^{-3}$) Pt(111)-surface $(\times)$ and with an ion bombarded Pt(111) surface at temperatures $T_s = 107\,\mathrm{K}$ ($\bullet$) and $T_s = 293\,\mathrm{K}$ ($\square$), respectively

essary to saturate a step atom, the density of step atoms is directly obtained. For instance, assuming that at saturation one CO molecule is bound per step atom, a step atom density of about $\Theta_d = 0.1$ is obtained from the position of the kink in curve (b) of Fig. 5.1. Or inversely, by knowing the density of step atoms (see Chap. 6), the average number of molecules bound at saturation at one step atom can be inferred.

The most efficient and generally applicable way to avoid the preferential binding of adsorbates at defect sites is to make them immobile immediately after adsorption by keeping the substrate at a sufficiently low temperature. The adsorbates then do not migrate to defect sites but remain randomly distributed as they arrived from the gas phase. Thus at sufficiently low coverages the adsorbates are isolated scatterers; the initial part of the slope of their adsorption curve supplies via (3.5) the correct cross-section for diffuse scattering. The method can be applied in principle to any adsorbate except for hydrogen atoms which appear to migrate also by tunneling [DiFoggio, Gomer, 1982] and thus cannot be kept immobile by cooling the substrate. The use of the method is illustrated in Fig. 5.1 for CO/Pt(111). By adsorbing CO on the same Pt(111) with defects, which resulted in curve (b) when the adsorption took place at 293 K, the black dots (•) are obtained when the Pt surface was kept at 107 K. The black dots fall exactly on the curve (a) obtained with the defect free Pt(111) at 293 K (in fact at any temperature) demonstrating that at 107 K the CO molecules, being immobile upon adsorption do not migrate to the defects but stay distributed randomly on lattice sites.

Now, to the main positive aspect of the binding of adsorbates at defect sites and the ensuing substantial cross-section overlap: a new method for the measurement of the surface migration of admolecules [Poelsema, Verheij, Comsa, 1982; 1983b]. A quick look at Fig. 5.1 suggests the method. Indeed, imagine that we adsorb CO on a Pt surface with a certain step density $\Theta_d$ at $T_s = 107$ K, i.e. we follow the black dots on curve (a). After some coverage $\Theta_{CO}$ is reached (preferably $\Theta_{CO} < \Theta_d$), the adsorption is interrupted by cutting the CO flow. Now, we start to increase the surface temperature, while $\Theta_{CO}$ stays obviously constant. As soon as the CO molecules become mobile, they will start to migrate landing eventually at step sites, where they are bound more strongly. Due to the substantial cross-section overlap the specular beam height increases: the signal moves vertically ($\Theta_{CO}$ constant) from curve (a) to (b). Thus the onset of CO migration is directly "visualized" and from the temperature at which the migration sets in, the activation energy for adsorbate diffusion is obtained.

The use of the method is illustrated in Fig. 5.2. The defects on the Pt(111) surface have been induced by ion bombardment and subsequent partial annealing. (The annealing has to be done at a temperature higher than the upper limit of the temperature range which has to be subsequently scanned when looking for the CO diffusion; this annealing is necessary in order to avoid substrate modification effects that interfere with the effect of CO diffusion itself.) The CO adsorption has been stopped at $\Theta_{CO} = 0.06$. Then the relative height $I/I_0$ is monitored during the linear increase of $T_s$; $\Theta_{CO}$ remains constant up to $T_s = 400$ K when desorption sets in. A drastic increase of the specular He intensity between 160 and 180 K

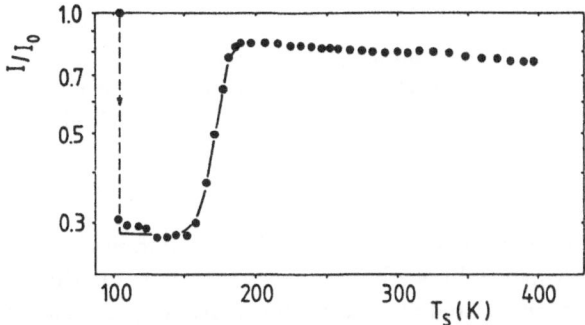

**Fig. 5.2.** Relative He specular peak height vs increasing surface temperature as measured on a Pt(111)-surface with defects on which CO has been preadsorbed ($\Theta_{CO} \approx 0.06$) at $T_s = 107$ K. The data taken with $E_{He} = 63$ meV and $\vartheta_i = 40°$ are corrected for Debye–Waller effects

is obvious from Fig. 5.2. This increase signals the onset of adsorbate mobility accompanied by a preferential occupation of defect sites. The solid curve is a fit of a simple hopping model to the experimental data. The fit yields an activation energy of 7 kcal/mole for diffusion of CO on Pt(111) [Poelsema, Verheij, Comsa, 1982]. This method to investigate surface diffusion permits measurements at low adsorbate concentrations, a regime hardly accessible to other techniques. (Only recently Reutt–Robey, Doren, Chabal and Christman [1988] have applied the same procedure using IR-detection; they were able to monitor even the kinetics of the CO migration.) In this coverage range mutual adsorbate interactions can be neglected; this facilitates a comparison with theories on adsorbate diffusion.

The result of an informative control experiment is shown in Fig. 5.3. In contrast to the case in Fig. 5.2, the Pt(111) surface has been prepared with the lowest possible defect density ($\Theta_d \approx 10^{-3}$). Therefore, only a few CO molecules can be accommodated on defect sites after becoming mobile; their number is so low that no measurable He intensity increase as the result of cross-section overlapping with defect sites should be expected. Moreover, since the CO molecules adsorbed on Pt(111) repel each other (see Sect. 4.3), they do not build islands and are expected to stay quasi-randomly distributed in the whole temperature

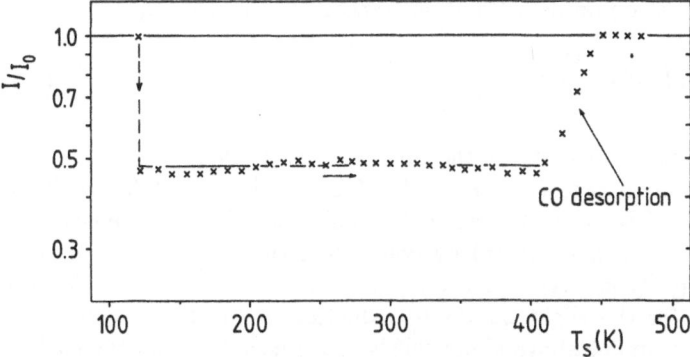

**Fig. 5.3.** Similar run as in Fig. 5.2 but on a "perfect" Pt(111) surface ($\Theta_d \approx 0.001$) (see text)

range. Indeed, after having adsorbed CO up to 0.04 at 120 K the subsequent temperature increase does not lead to any intensity increase until CO desorption occurs at temperatures above 400 K (see also Chap. 8). Consequently, the control-experiment confirms firstly the very low defect concentration on the Pt(111) surface, and secondly the absence of CO island formation on Pt(111) which has been previously stated to occur [Crossley, King, 1980], see also [Greenler et al., 1985].

The application of TEAS to measure adsorbate diffusion, as described above, is based on its capability to distinguish between adsorbates randomly distributed on defect free sites and those concentrated on defect sites. Advantage is taken of the stronger binding of the adsorbates on defect sites with respect to defect free terrace sites.

In the case of thermal equilibrium between the gas phase and the adsorbed phase the concentration of adsorbates on sites of type $a$ can be written as (see, e.g., Ibach, Erley, Wagner [1980])

$$\Theta_a = \frac{n_a p \gamma_a \exp(E_a/kT_s)}{1 + p \gamma_a \exp(E_a/kT_s)} \quad , \tag{5.1}$$

where $n_a$ denotes the ratio of the number of sites of type $a$ to the total number of surface atoms, $E_a$ the binding energy on these sites, $\gamma_a$ a site specific factor containing the translational, rotational and vibrational partition functions of the adsorbate and $p$ the pressure of the adsorbate in the 3D-gas phase: $\Theta_a = 0$ for $p = 0$ and $n_a^{-1}\Theta_a = 1$ for $p = \infty$. The quantities associated with defect and terrace sites are indicated with subscripts d and t, respectively. In the limit of small concentrations, i.e. $n_t^{-1}\Theta_t$ and $n_d^{-1}\Theta_d \ll 1$ (i.e. $p\gamma_{t,d} \exp(E_{t,d}/kT_s) \ll 1$), the adsorbate concentrations are in good approximation given by

$$\Theta_d = p n_d \gamma_d \exp\left(\frac{E_d}{kT_s}\right)$$

and

$$\Theta_t = p n_t \gamma_t \exp\left(\frac{E_t}{kT_s}\right) \quad . \tag{5.2}$$

The relative concentration on defect and on terrace sites is given by

$$\frac{\Theta_d}{\Theta_t} = \frac{n_d \gamma_d}{n_t \gamma_t} \exp\left(\frac{E_d - E_t}{kT_s}\right) \quad , \tag{5.3}$$

provided that thermal equilibrium *on* the surface is established.

In the most common case (illustrated for CO/Pt above and for H/Pt in Chap. 8) $\Delta E = E_d - E_t > 0$. According to (5.3), the relative population of terrace and defect sites can be varied by controlling the surface temperature; more specifically, $\Theta_t/\Theta_d$ increases with temperature for $\Delta E > 0$. This feature may be responsible for the slight but finite reduction of the He reflectivity in the reversible curve segment above about 200 K (see Fig. 5.2). From the surface temperature dependence information on the difference in binding energy $\Delta E$ can

44

be deduced. Unfortunately, this appears to be hardly possible for CO on Pt(111) since it desorbs already below the surface temperature at which $\Theta_t$ becomes significant.

For the case of Pb/Cu(001) Sánchez, Ibañez, Miranda and Ferrer [1986] found a behaviour that resembles the CO/Pt(111) case. The adsorption curves, taken at $T_s = 345$ and 465 K, as shown in their Fig. 2, are similar to curve ($b$) in Fig. 5.1. In this particular case, it is possible to determine $\Delta E$, since a significant amount of the adsorbed Pb atoms populates terrace sites well below their desorption temperature. This conclusion is drawn from the reversible behaviour of $I/I_0$ as a function of the surface temperature, at a constant Pb coverage. This behaviour is exemplified in Fig. 5.4 in which the effective cross-section, defined as $\Sigma' = (1 - I/I_0)/(n_s\Theta)$, is plotted vs $T_s$ at constant $\Theta_{Pb} = 0.003$. Small $\Sigma'$ values indicate substantial overlap between the cross-sections of the Pb adatoms and the steps; i.e. steps are preferentially occupied at relatively low $T_s$. At temperatures above 500 K the concentration of Pb atoms on terrace sites equals approximately that on defect sites. From this behaviour a binding energy difference $\Delta E = 0.4 \pm 0.1$ eV is inferred.

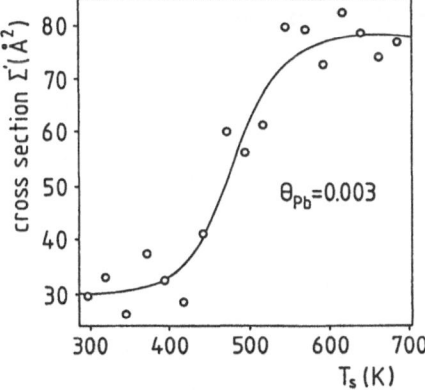

**Fig. 5.4.** Effective (see text) cross-section $\Sigma'$ for diffuse scattering of He on Pb adsorbed on Cu(001) as a function of the substrate temperature $T_s$; $E_{He} = 63\,meV$, $\vartheta_i = 75°$ and $\Theta_{Pb} = 0.003$ (after Sanchez, Ibañez, Miranda, Ferrer [1986])

It is interesting to note that the dependence of $I/I_0$ on $\Theta_{Pb}$ can also be described in a way similar to the CO/Pt(111) case. The adsorption curve follows closely the modified lattice gas expression (3.8) with m = 8/3 and $\Sigma = 78\text{Å}^2$.

## 5.2 Coadsorption

Another widespread kind of heterogeneous systems is that in which different species of atoms or molecules are coadsorbed. They have fundamental and practical importance. It will be shown that both attractive and repulsive interactions between the coadsorbates influence deeply their distribution on the surface.

### 5.2.1 Xe/CO/Pt(111) — Attractive Interaction

The shape of the Xe adsorption curve ($I/I_0$ vs Xe exposure) depends dramatically on the presence of preadsorbed CO. This is illustrated in Fig. 5.5 for Xe adsorption on a defect-free Pt(111) surface at $T_s = 90$ K for two CO precoverages ($\Theta_{CO} = 0.038$ and $0.077$, respectively). Two substantial differences with respect to the Xe adsorption curve on the clean (not CO-precovered) Pt(111) surface are noted (compare Fig. 4.6): the initial slope is considerably smaller and 2D-condensation sets in at larger Xe exposures. Both effects can be understood by assuming attractive interactions between CO and Xe, leading to the formation of Xe-CO clusters. As a consequence the scattering cross-sections of the newly arriving Xe atoms overlap with those of the preadsorbed CO molecules. This leads, as argued before, to a reduction of the effective scattering cross-section, and consequently to a smaller initial slope of the adsorption curve. On the other hand, the Xe atoms participating in the Xe-CO clusters are withdrawn from the 2D-Xe gas phase. Thus in order to reach the 2D-Xe vapour pressure a correspondingly higher exposure is required. The experimental result is consistent with the well-known fact that small amounts of impurities have a substantial influence on 2D gas-solid phase transitions, see e.g. [Glachant, Bardi, 1979]; the analysis of the He scattering data via the overlap approach supplies in addition a pertinent picture for the actual mechanism of this influence.

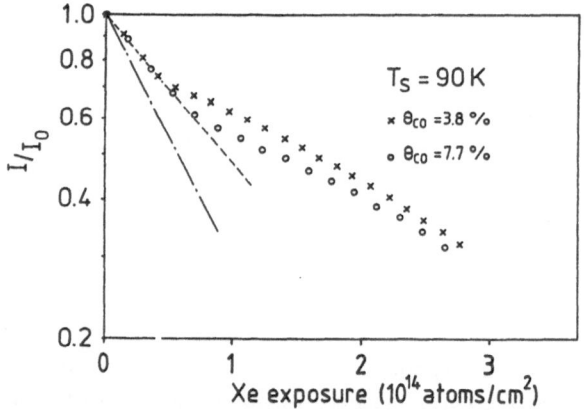

**Fig. 5.5.** Relative He specular peak height as a function of Xe exposure on a Pt(111) surface precovered with CO; $\Theta_{CO} = 0.038$ (×) and $\Theta_{CO} = 0.077$ (o), respectively. The data were taken at $T_s = 90$ K with an 63 meV He beam incident at $\vartheta_i = 40°$. The dash-dotted curve shows the lattice gas behaviour of Xe on a clean Pt(111) surface below the 2D-condensation coverage. The dashed curve emphasizes the common initial part of the Xe adsorption curves on CO-precovered surfaces

### 5.2.2  H/CO/Pt(111) — Repulsive Interaction

The procedure used in this section to explore the influence of the mutual interaction between H and CO on their lateral distributions during coadsorption is based on the radically different nature of He scattering from these two adsorbates [Bernasek, Lenz, Poelsema, Comsa, 1987; Lenz, Poelsema, Bernasek, Comsa, 1988]. Indeed, according to the definition in Sect. 3.5.1, CO is a perfectly diffuse scatterer at least for $T_s > 80$ K: the specular He intensity scattered from a CO-saturated surface area is less than $10^{-2}$ of the specular He intensity scattered from the same *clean* area. In contrast, H is a (highly) reflecting adsorbate (Sect. 3.5.2): the specular He intensity scattered from a *H-saturated* surface area is only $\varrho = 0.7$, (respectively $\varrho = 0.5$) times lower than the specular He intensity scattered from the same *clean* area at $T_s = 80$ K (respectively $T_s = 150$ K). Both values correspond to a He beam of 63 meV incident at $\vartheta_i = 40°$. In other words, while the mirror-like clean Pt(111) surface is made completely "black" by CO adsorption, the H-saturated surface areas are still reflecting mirror-like with a reflectivity reduced only by a factor of 0.7 (and 0.5, respectively). Details of the scattering from H-saturated surfaces are discussed in Sect. 8.1.1. The procedure is certainly not restricted to the H/CO system, but is applicable in all cases in which the scattering behaviour of the partners is so different. The actual lateral distribution of the system H/CO discussed here is particularly interesting for the detailed understanding of its reaction kinetics.

The procedure is schematically illustrated by means of Fig. 5.6. The clean Pt(111) surface is first exposed to CO, until a certain precoverage $\Theta_{CO}$ is reached (Fig. 5.6 left). As shown in Sect. 4.3 the distribution of CO corresponds to that of a lattice gas with non-occupied nearest neighbour sites; the reflectivity is given by (4.5) as $r(\Theta_{CO}, \Theta_H = 0) = I/I_0 = (1 - 3\Theta_{CO})^{n_s \Sigma_{CO}/3}$. The cross-section $\Sigma_{CO}$ of each molecule is represented by a hatched circle and the size of the black circles corresponds to close packed spheres at the maximum coverage $\Theta_{CO} = 0.58$ of CO on Pt(111). Thus the reflectivity is equal to the squared fraction of the surface non-covered by the hatched circles, i.e. the white area, for which the local reflectivity is unity.

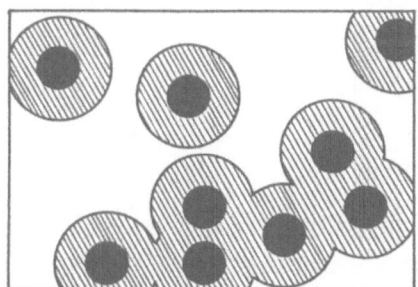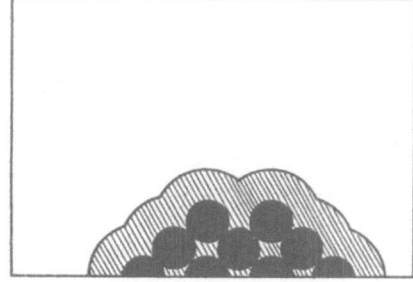

**Fig. 5.6.** Schematic representation of the same average density of CO molecules in the H/CO /Pt(111) coadsorption system for a random distribution (*left*) and for segregation of the molecules into islands (*right*); the He atoms are scattered diffusely from the hatched areas and specularly from the white ones

In the second step, the Pt surface partially precovered with CO is exposed to hydrogen up to saturation. Let us now assume that the coadsorption of hydrogen does not change the distribution of the preadsorbed CO and ask ourselves what will be the reflectivity of the H-saturated surface. We have to make assumptions concerning the He scattering from CO and H when these are coadsorbed. The simplest assumption is that the scattering behaviour is essentially not modified by coadsorption, i.e. (a) the size of the scattering cross-section for He diffuse scattering of the CO molecules ($\Sigma_{CO}$) is not changed by the presence of H atoms and (b) the local reflectivity on the white area (non-covered by the hatched circles) has the same value $\varrho$ as that of the H-saturated Pt surface in the absence of CO. Thus, if upon H saturation the CO distribution remains as sketched in Fig. 5.6 (left) the expected reflectivity will be simply

$$r(\Theta_{CO}, \Theta_H = \text{sat}) = \varrho(1 - 3\Theta_{CO})^{n_s \Sigma_{CO}/3} \quad , \qquad (5.4)$$

i.e. slightly less ($\varrho = 0.5$ at $T_s = 150\,\text{K}$) than before H adsorption. The experimental result (Fig. 5.7 solid curve) is at odds with this expectation. Indeed the reflectivity of the saturated coadsorption system (thick solid line segment at high $H_2$-exposure) is about 33 times *larger* than before $H_2$-exposure (thick solid line at the left): $r(\Theta_{CO} = 0.18, \Theta_H = \text{sat}) = 33r(\Theta_{CO} = 0.18, \Theta_H = 0)$. If we maintain the two assumptions concerning the He scattering made above (which will be shown below to be correct), the only other possibility is that the adsorption of hydrogen changes radically the distribution of the preadsorbed CO. The dramatic enhancement of the He reflectivity indicates that the cross-sections of the preadsorbed CO are strongly overlapping upon hydrogen adsorption; i.e., the CO molecules become confined to islands of high local density. This conclusion has been confirmed unambiguously by low-energy electron diffraction (LEED). Upon H saturation a $c(4 \times 2)$-LEED pattern is observed in a broad range of CO

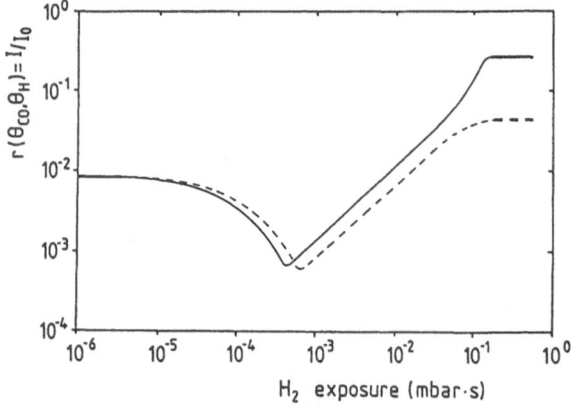

**Fig. 5.7.** Relative He specular peak height as a function of $H_2$ exposure for a CO precoverage of $\Theta_{CO} = 0.18$ on Pt(111) at different substrate temperatures: $T_s = 80\,\text{K}$ (*dashed curve*) and $T_s = 180\,\text{K}$ (*solid curve*). The He reflectivities for the H-free and H-saturated adlayers are emphasized by thicker horizontal segments. ($E_{He} = 63\,\text{meV}$ and $\vartheta_i = 40°$)

precoverages ($\Theta_{CO} = 0.05 - 0.30$). This pattern is known to appear only in a narrow range of CO coverages on Pt(111) around $\Theta_{CO} = 0.50$, see e.g.[Steininger, 1982]. Thus the hydrogen compels the preadsorbed CO molecules to segregate into islands of $\Theta^l_{CO} = 0.50$ local density, *independent* of the amount of CO precoverage.

In order to check assumption (b) (i.e. that the local reflectivity $\varrho$ of the H-saturated areas is not changed by CO adsorption) the reflectivity of the H-saturated coadsorbate layer $r(\Theta_{CO}, \Theta_H = \text{sat})$, normalized to that of the CO-free layer $r(\Theta_{CO} = 0, \Theta_H = \text{sat}) = \varrho$, has been plotted as a function of CO precoverage at $T_s = 180$ K in Fig. 5.8. The local density in the CO islands being $\Theta^l_{CO} = 0.50$, the "size" of a CO-molecule within an island is $U = (\Theta^l_{CO} n_s)^{-1}$, where $n_s$ is the density of the Pt atoms on the Pt(111) surface. Thus upon compression the CO molecules will occupy a fraction $U n_s \Theta_{CO} = 2\Theta_{CO}$ of the total area. Assuming now that (b) holds and neglecting island-edge effects, the reflectivity of the H-saturated coadsorbate layer would be $r(\Theta_{CO}, \Theta_H = \text{sat}) = \varrho(1 - U n_s \Theta_{CO}) = \varrho(1 - 2\Theta_{CO})$. Thus the data in Fig. 5.8 should be on the straight line

$$\frac{r(\Theta_{CO}, \Theta_H = \text{sat})}{r(\Theta_{CO} = 0, \Theta_H = \text{sat})} = \frac{\varrho(1 - U n_s \Theta_{CO})}{\varrho} = 1 - 2\Theta_{CO} \tag{5.5}$$

plotted as solid line in Fig. 5.8. The agreement with the experiment confirms assumption (b). The slight deviation at higher coverages is probably due to finite island size effects, i.e., diffuse scattering from the edges of the H islands.

Note that the island formation takes place in a two-component system in which the adsorbate-absorbate interactions in the single-component systems are purely repulsive (island formation is commonly connected with attractive interactions). This may imply that the CO-H repulsion is stronger than the average of the CO-CO and the H-H repulsion.

Let us now determine the cross-section for diffuse scattering of He on isolated CO molecules in the coadsorbate system at H saturation, in order to check assumption (a) above. We have thus to expose the CO-precovered surface to hydrogen as above, but at a surface temperature low enough for at least one of

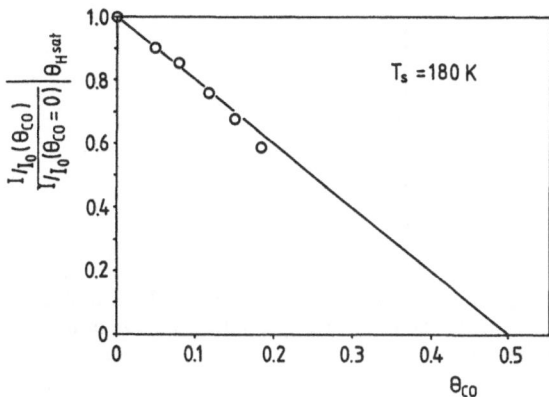

**Fig. 5.8.** Normalized He reflectivity ($r = I/I_0$) of the H/CO-coadsorbed layer at H saturation as a function of CO precoverage. The data were taken with a 63 meV He beam incident at $\vartheta_i = 40°$ on Pt (111); the temperature during H adsorption was $T_s = 180$ K (see text)

the components to be immobile. The temperature was $T_s = 80\,$K, so that according to the experiments shown in Sect. 5.5, CO is expected to be immobile on Pt(111). (The hydrogen atoms are certainly mobile down to much lower temperatures [DiFoggio, Gomer, 1982].) From a number of H-exposure curves such as those in Fig. 5.7 (but now at 80 K) taken at various CO precoverages, the data in Fig. 5.9 are obtained. As in Fig. 5.8 the normalized reflectivity $r(\Theta_{CO}, \Theta_H = $ sat$)/r(\Theta_{CO} = 0, \Theta_H = $ sat$)$ is plotted vs CO precoverage. The dashed curve represents, according to (5.4), the normalized reflectivity for the lattice gas like distributed CO (immobile during and after H adsorption), assuming that both (a) and (b) are correct ( (a) $\Sigma_{CO}^{He}(\Theta_H = $ sat$) = \Sigma_{CO}^{He}(\Theta_H = 0)$ and (b) the total He reflectivity from the H-saturated layer remains $\varrho = r(\Theta_{CO} = 0, \Theta_H = $ sat$)$ for $\Theta_{CO} \neq 0$). Assumption (b) has been already shown to be correct. Up to $\Theta_{CO} \approx 0.05$ the data lie on the dashed curve showing that assumption (a) is also correct. The deviation of the data above $\Theta_{CO} = 0.05$ towards larger He reflectivities indicates that in the presence of $\Theta_H = $ sat, there is, even at $T_s = 90\,$K, a certain tendency towards CO clustering.

The reflectivity curves in Figs. 5.8 and 5.9 bear a certain similarity with those in Fig. 5.1. In both cases, the cross-section overlap is kept low (and thus also the reflectivity) at low temperatures, in contrast to the high temperatures where the overlap (and the reflectivity) are larger due to CO island formation or to binding at defects, respectively. The low overlap situation corresponds to the lattice gas like distribution of CO kept in a metastable state by the immobility of the CO molecules at these low temperatures. Thus in analogy with the procedure demonstrated in Sect. 5.1 the onset of CO mobility can be directly visualized also in the coadsorbate system. This is shown in Fig. 5.10. The Pt(111) surface has been precovered with CO up to $\Theta_{CO} = 0.025$ and then exposed to hydrogen at $T_s = 87\,$K up to saturation. The temperature of the surface prepared in

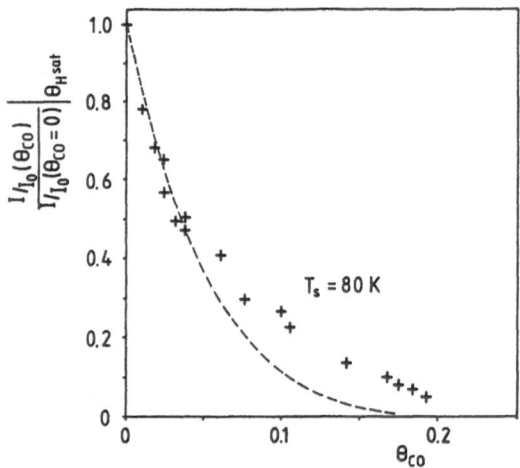

**Fig. 5.9.** Same experiment as in Fig. 5.8, but for H adsorption at $T_s = 80\,$K. The dashed curve corresponds to the behaviour expected for the CO lattice gas with exclusion of nearest neighbours (4.5) which fits the data in Fig. 4.2

50

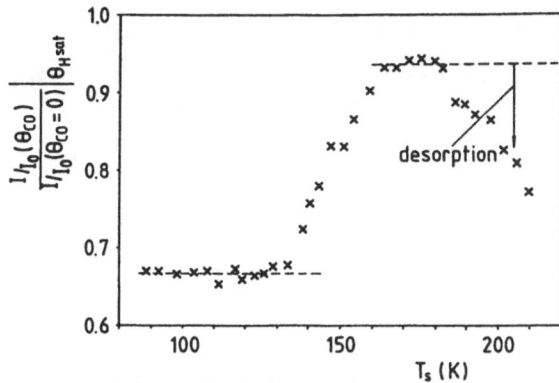

**Fig. 5.10.** Normalized He reflectivity of a H-saturated H/CO layer prepared at $T_s = 87\,$K (CO precoverage $\Theta_{CO} = 0.025$) during surface temperature increase (compare Figs. 5.8 and 9). The He reflectivity decrease above $T_s \approx 180\,$K is due to $H_2$ desorption. The data measured with $E_{He} = 63\,$meV, $\vartheta_i = 40°$ are corrected for Debye–Waller effects

this way is increased linearly while monitoring the He reflectivity. A comparison between Figs. 5.10 and 5.2 shows that in the presence of $\Theta_H = $ sat the onset of the CO mobility is at about $20 - 30\,$K lower temperature. This lowering of the activation energy for CO migration is probably connected with the reduction of the CO binding energy on the surface, due to the strong repulsive interaction with hydrogen. (Activation barriers for surface diffusion are known to be roughly proportional to the binding energies.) Another difference with respect to the $\Theta_H = 0$ case in Fig. 5.2 is the marked decrease in He reflectivity above $180\,$K. This is due to the desorption of hydrogen. The He reflectivity recovers eventually — upon hydrogen desorption — the value corresponding to the $\Theta_{CO} = 0.025$ precoverage, demonstrating that during H adsorption and the following heat treatment no CO is lost from the surface.

# 6. The Characterization of Stepped Surfaces by Means of Coherent and Diffuse Helium Scattering

## 6.1 Introduction

Any real monocrystalline surface contains a non-zero concentration of atomic steps. From this point of view the real surfaces may be classified into two categories: nominally (regularly) and "randomly" (irregularly) stepped surfaces.

Nominally stepped surfaces are defined as surfaces of monocrystals consisting of terraces with a low Miller index [(001), (011) or (111)] separated by monatomic steps. If the step rows are not oriented along one of the main crystal axes they possess kinks and the surfaces are called regularly kinked surfaces. Nominally stepped surfaces are obtained by cutting a monocrystalline rod along a plane with the normal deviating by an angle $\alpha$ from the normal to a low index plane (see also Chap. 2). The average terrace width is given by $\overline{\Delta} = h/\tan\alpha$, where $h$ denotes the step height, i.e., the interlayer distance of the low indexed terrace planes. After UHV high temperature treatment (see also Chap. 2) these surfaces turn out to have a nearly periodic structure, i.e., the width of the step-step distance distribution is relatively narrow. This is attributed to repulsive interactions between neighbouring steps, see e.g. [Besocke, Krahl–Urban, Wagner, 1977]. Nominally stepped surfaces can therefore be characterized as a regular array of terraces, separated by equidistant monatomic steps in a monotonic step-up (or step-down) sequence.

A concise but also comprehensive definition of the randomly stepped surfaces — the actual subject of the present discussion — can hardly be given. Definitions like "non-nominally stepped surfaces" or "unintentionally stepped surfaces" are neither helpful nor even comprehensive. We will thus just analyse the various origins of this type of surface. The conclusion will be that the most appropriate characterization of the so-called randomly stepped surfaces is that that they are random on a macroscopic scale but regularly stepped on microscopic areas: the actual size of macroscopic and microscopic scales may vary within wide limits. The following causes for the build up of this type of surface have to be discussed.

Let us first consider the ideal case of a perfect preparation technique, i.e., the crystal can be oriented, cut and polished along an arbitrary, perfect mathematical plane. However, even the best available "monocrystalline" rods have a mosaic structure; the rod consists of crystallites with slightly differing orientations. Thus in principle the perfect cutting plane can be chosen to coincide with a low index plane of only one of the crystallites. The surfaces cut on all other crystallites will be regularly stepped surfaces; the macroscopic surface is therefore composed of small regularly stepped surfaces each characterized by an angle $\alpha_i$. Typical differences in mosaic orientations as determined by means of X-ray diffraction topography amount to $0.025 - 0.1°$ for Pt and $0.15 - 0.5°$ for Ni and Cu. The mosaics have typical dimensions of $10 - 100 \,\mu m$ [Linke, 1988].

The second idealized case is that of a perfect monocrystalline rod. The careful use of the state-of-the-art orientation and cutting procedures results in a surface oriented within $\alpha_m = \pm 0.1°$ (at best $\pm 0.05°$) with respect to a given plane of the crystal rod. If this plane is a low index one, the step density will still be $\sim 0.002$ (at best $0.001$). Moreover, even the most careful polishing of the spark cut surface results in some waviness, i.e., in a variation of $\alpha$ along the surface. High temperature annealing reduces most of this waviness by surface diffusion [Bonzel, 1983], but annealing of long wave roughness requires exceedingly long times. Note, however, that neither the $\alpha_i$ distribution, due to the mosaic structure of the rod, nor the misalignment $\alpha_m$, due to the limited precision of the orientation and cutting procedures, can be "annealed out".

The third case includes the "intentionally" randomly stepped surfaces and represents the final goal of this chapter. The steps are known to play an important role in surface reactions [Somorjai, 1981]. This has been demonstrated in countless investigations performed on various surfaces with different step densities. All of these surfaces have been obtained by cutting the crystal surface at a desired angle $\alpha$ with respect to that of the low index planes of the crystal. Thus for each step density a newly cut crystal had to be used. This is not only time and work consuming, but makes accurate and reproducible comparisons between surfaces of different step densities hardly possible: each step density corresponds to a new crystal (at least to a newly cut crystal) with a different history and thus with impurities and other types of defects differing in nature and concentration. The procedure which will be presented here in Sect. 6.5.3 allows one to vary $in\ situ$ the step density of the $same$ crystal surface by more than two orders of magnitude ($0.001 < \Theta_d < 0.15$). This is achieved by ion bombarding the surface at a fixed, relatively high surface temperature. The procedure could be developed and implemented by using the unique characterization capabilities of He scattering $during$ ion bombardment, which will be discussed in detail in this chapter. It has been shown [Mullins, 1959; Cherns, 1977; Osakabe, Tanishiro, Yagi, Honjo, 1981; Verheij, Van den Berg, Armour, 1982; Verheij, 1982] that the stepped surfaces produced in this way are also not arbitrarily random; the ledges separating the low index terraces are oriented preferentially along low indexed (close packed) directions [Lahee, Manson, Toennies, Wöll, 1986]. Thus also in this case, the surfaces exhibit some order at a microscopic scale and are only macroscopically disordered. Note that, in contrast to the other two cases above, these "intentional" steps can readily be annealed out by high temperature treatment.

The characterization of randomly stepped surfaces by He and electron scattering has been discussed in a number of papers in recent years [Henzler, 1970; 1978; 1982; Cowley, Shuman, 1973; Schulze, Henzler, 1978; Welkie, Lagally, 1979; Lapujoulade, 1981; Lu, Lagally, 1982; Levi, Spadacini, Tommei, 1982; Spadacini, Tommei, 1983]. These theories are rather complicated and not very intuitive. We prefer to use here a more intuitive approach, which is based on the conclusion reached above: the randomly stepped surfaces represent a random ensemble of nominally stepped microscopic surfaces. It has been demonstrated that, starting with a given scattering pattern, the different approaches lead to identical step densities; differences appear only with respect to the distribution of the terrace widths [Verheij, Poelsema, Comsa, 1985].

The outline of this chapter is as follows. First the He diffraction from regularly stepped surfaces is described and exemplified. We show then how the step density of a randomly stepped surface is estimated by monitoring the specular He peak height as a function of perpendicular momentum exchange. This is followed by the discussion of the detailed He peak shape analysis which leads to the determination of the size of the cross-section for He diffuse scattering from atomic steps. These procedures are then applied to the study of the surface morphology during ion bombardment as a function of temperature and during crystal growth. Finally, the procedure to create in situ surfaces with desired step densities is emphasized.

## 6.2  Helium Diffraction from Regularly Stepped Surfaces

We consider first a regularly stepped surface consisting of (111)-terraces, separated by monatomic steps. The plane of incidence of the helium beam is chosen perpendicular to the steps. This situation is sketched schematically in Fig. 6.1. The angle between the normal to the (111)-terraces and that to the macroscopic surface plane is denoted by $\alpha$; the height difference between two adjacent terraces by $h$. He diffraction from a nominally stepped surface is particularly simple in the case of close packed (111)-terraces of metal fcc-lattices. The terraces reflect the helium beam like a mirror (Chap. 3). Therefore, the diffraction pattern is analogous to the well-known case of light scattering form an echelette grating [Comsa, Mechtersheimer, Poelsema, Tomoda, 1979; Sommerfeld, 1959]. Qualitatively, most of the scattered intensity is concentrated around the specular direction with respect to the terraces.

Thus one may consider that the He diffraction takes place on a periodic lattice with the terrace mirror as unit cell. The diffraction is described by the product of a structure function and a form factor function. The form factor is determined by scattering from a unit cell. In the plane wave approximation the amplitude is identical to the slit function known from optics

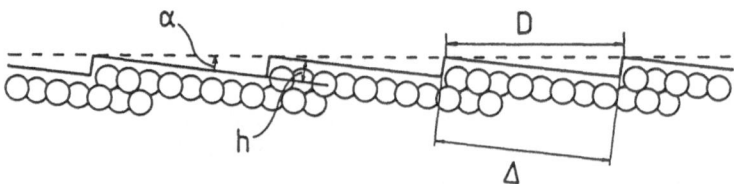

**Fig. 6.1.** Sketch of a regularly stepped fcc(997) surface. The plane of the drawing is perpendicular to the (1$\bar{1}$1) oriented steps. The dashed line indicates the macroscopic surface plane. For explanation of symbols see text

$$A/A_0 = \frac{1}{i\varphi}\left(e^{i\varphi} - 1\right) \tag{6.1}$$

with

$$\varphi = \frac{2\pi D}{\lambda}\left(\sin\vartheta_i - \sin\vartheta_f\right) \quad \text{and} \quad D = (h^2 + \Delta^2)^{1/2} \tag{6.2}$$

where $\lambda$ and $\Delta$ denote the He wavelength and the terrace width, respectively; $\vartheta_i$ and $\vartheta_f$ are incident and exit angle with respect to the terrace normal. The corresponding intensity, peaked at $\vartheta_f = \vartheta_i$, is

$$I = A^* \cdot A = A_0^2 \frac{\sin^2(\varphi/2)}{(\varphi/2)^2} \quad . \tag{6.3}$$

The FWHM of this form factor, having a maximum in the specular direction to the terraces, amounts to

$$\Delta\vartheta_{1/2} = \frac{\lambda}{D} \cdot \frac{1}{\cos\vartheta_i} \quad . \tag{6.4}$$

The structure function for a perfectly ordered, stepped surface of infinite size is given by

$$I = \delta\left(\sin\vartheta'_i - \sin\vartheta'_f - n\frac{\lambda}{D}\right) \tag{6.5}$$

in which $\vartheta'_f = \vartheta_f - \alpha$ and $\vartheta'_i = \vartheta_i + \alpha$ denote the angles of incidence and the exit angle, respectively, defined with respect to the *macroscopic* surface normal. $\delta$ represents a delta function and $n$ the order of the diffraction peaks. For facets considerably larger than the transfer width (see Chap. 2) the FWHM values of the Bragg peaks (6.5) are small with respect to instrumental widths, allowing the use of the $\delta$ representation of (6.5).

The total diffraction pattern consists of the product of (6.3) and (6.5): the height of the $\delta$ peaks is determined by expression (6.3); expression (6.3) is the envelope of the $\delta$ peaks. For each angle of incidence $\vartheta'_i$ the positions of the diffracted beams are given by the structure function of expression (6.5). Variation of the angle of incidence leads to a continuous shift of the positions of the diffracted beams with respect to the specular direction to the terraces ($\vartheta_f = \vartheta_i$). Since the specular directions to the terraces and the macroscopic one are separated by an angle $2\alpha$, the diffracted beams of a certain order $n$ can be made to coincide with the specular direction to the terraces, i.e. with the maximum of the form factor (6.3), by a suitable choice of the angle of incidence $\vartheta'_i$. In this situation, illustrated in Fig. 6.2 for $n = 6$ (solid line), the adjacent terraces scatter in-phase. Then, the path length difference $\Delta p$ for He scattering from adjacent terraces, given by

$$\Delta p = h(\cos\vartheta_i + \cos\vartheta_f) \tag{6.6}$$

equals $\Delta p = n\lambda$ (for the example in Fig. 6.2: $n = 6$). The diffraction peaks of order $n-1$ and $n+1$ ($n = 5$ and $n = 7$) are positioned in the first minima of the form function (compare Fig. 6.2). Thus the total diffraction pattern for in-phase

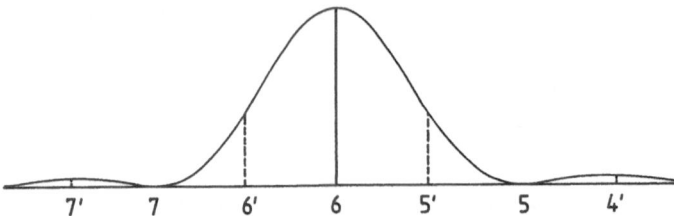

**Fig. 6.2.** He diffraction pattern for a regularly stepped surface. The solid curve shows the structure function (6.3). The vertical lines and the numbers indicate the positions and the order of the Bragg peaks (lattice function, (6.5)) for the case that two neighbouring terraces reflect in-phase (solid lines, $n$) and out-of-phase (dashed lines, $n'$)

scattering consists of a single $\delta$ peak, located specular to the terraces regardless of the terrace width. (We have neglected step edge effects.)

For another choice of the angle of incidence $\vartheta_i'$ the diffraction peaks of, e.g., order $n-1$ and $n$ can be made to fall on each side of the maximum of the amplitude function (6.3), i.e., of the specular direction to the terraces. For anti-phase scattering, i.e. $\Delta p = (n - \frac{1}{2})\lambda$ [see (6.6)], one observes two main peaks (dashed lines for $n = 5'$ and $n = 6'$ in the example of Fig. 6.2 for $\Delta p = 5.5\lambda$) around the specular direction to the terraces at a mutual distance of $\Delta\vartheta_f = \lambda/[D\cos(\vartheta_i - 2\alpha)]$. Therefore, in contrast to the in-phase scattering situation one observes two peaks instead of one, with a separation depending on the facet angle $\alpha$. Moreover, one observes also minor peaks at the loci of the higher order maxima of the unit cell function (see Fig. 6.2, dashed lines for $n = 4'$ and $n = 7'$).

Figure 6.3 shows an experimental diffraction pattern ($\times$) obtained with a 63 meV He beam incident at $\vartheta_i = 52°$ in the step down direction on a Pt(997) surface. The solid curve is obtained by applying (6.3) and (6.5) and accounting for the finite angular and energy resolution of the instrument [Comsa, Mechter-sheimer, Poelsema, Tomoda, 1979]. The calculated curve matches the experimental data, supporting the validity of the approach outlined above.

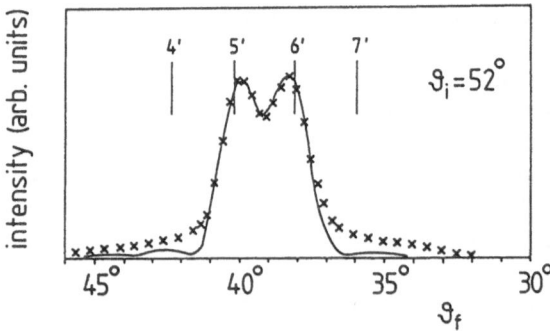

**Fig. 6.3.** Experimental diffraction pattern obtained with 63 meV He beam incident at $\vartheta_i = 52°$ on Pt(997) in the step-down direction ([7, 7, −18] azimuth). The solid curve is calculated according to (6.3 and 5) by taking the instrumental broadening into account [Comsa, Mechtersheimer, Poelsema, Tomoda, 1979]

## 6.3 Helium Diffraction from Randomly Stepped Surfaces; a Qualitative Description

The discussion on the origin of randomly stepped surfaces in Sect. 6.1 has shown that a natural way to rationalize these surfaces is to consider that they consist of an ensemble of regularly stepped surfaces (facets). Each facet is characterized by the angle $\alpha_i$. The group of facets $\alpha_i - \Delta\alpha < \alpha < \alpha_i + \Delta\alpha$ cover a certain fraction $\Delta s/s$ of the total area. The total diffraction pattern is obtained by a summation of single facet diffraction patterns (like that shown in Fig. 6.2), weighted by the corresponding fractional area $\Delta s/s$. The procedure is described in detail by Verheij, Lux and Poelsema [1984]. Here, we discuss only the qualitative aspects of the results.

Under in-phase conditions, each facet contributes with a unique peak oriented in the specular direction with respect to the terrace plane. Therefore, if the mosaic structure can be neglected, the sum-pattern will be a $\delta$ peak, in principle identical to that from an ideal Pt(111) surface. (As shown in the next section, the actual peak intensity from a randomly stepped surface is lower due to the diffuse scattering at the step edges). This is the consequence of constructive interference of *all* facets regardless of their facet angles. For destructive interference, i.e. anti-phase conditions, the situation differs dramatically. Depending on the orientation of their corresponding facets, the two "symmetric" peaks (see Fig. 6.2) and their smaller companions have different positions with respect to the specular direction. Consequently, for a distribution of facets one expects a broadened peak centered around the specular direction to the terraces. In a first approximation, the width of this peak will be proportional to the average value $\overline{|\alpha|}$ of the facet angles and thus inversely proportional to the mean terrace width $\overline{\Delta}$, see (6.4). A measurement of the width of the anti-phase peak is the simplest way to estimate the mean terrace width, i.e. the step density [Lapujoulade, 1981]. A more precise determination of the mean terrace width and even an estimation of the terrace width distribution is feasible only upon a detailed analysis of the peak shape [Verheij, Poelsema, Comsa, 1985].

The sensitivity in measuring the peak-broadening under anti-phase conditions, i.e. ultimately in measuring step densities, is limited by the instrumental resolution. The specular peaks measured under in-phase conditions are not $\delta$ peaks, but have a certain width $\Delta\vartheta_f \neq 0$. With some effort, a broadening under anti-phase conditions larger than about $0.1\Delta\vartheta_f$ may be estimated. The instrumental width of the specular beam is correlated with the transfer width $w$ (2.1). Thus the largest average terrace widths which can be detected are about $10w$. For example, with an instrument with a total angular resolution of $0.2°$ average terrace widths up to about 2500 Å can be estimated.

The measurement of peak profiles with reasonable accuracy is a rather time consuming procedure. This is certainly necessary when the terrace width distribution is sought by means of a detailed analysis. In most cases, however, only an estimation of the average terrace width, i.e. the step density, is needed. As shown above, this can be directly obtained from the knowledge of the peak width under in- and anti-phase conditions. For estimating the peak width it is not necessary to monitor the whole peak profile: it is sufficient to measure the peak height. Indeed, the whole peak intensity (i.e. the peak area or more exactly its volume) is largely the same for in-phase and anti-phase scattering. The peak height is thus a fair measure of the peak width. Typical experimental data are shown in Fig. 6.4 for two (111)-faces prepared from the same Pt crystal [Poelsema, Palmer, Mechtersheimer, Comsa, 1982]: ($a$) has been oriented with a conventional procedure with an accuracy of $|\alpha| = 0.5°$ and ($b$) with a much improved accuracy $|\alpha| < 0.1°$. The height of the specular peak is plotted as a function of the exit angle $\vartheta_f (= \vartheta_i)$. Curve ($a$) shows an oscillatory behaviour. The minima and maxima indicate anti-phase ($\downarrow$)and in-phase ($\uparrow$) scattering respectively. The positions of the extrema correspond to an interlayer spacing $h = 2.27 \pm 0.05$ Å in agreement with the spacing between (111)-layers in a Pt crystal. A rough evaluation of the amplitude of the oscillations yields a mean terrace width of about $\overline{\Delta} = w/2$. Curve ($b$) shows no oscillations at all; the increase of the peak height with increasing angle of incidence follows closely the Debye–Waller behaviour. The absence of oscillations in the latter curve implies that the in- and anti-phase peak widths are indistinguishable. Thus the average terrace width has to be more than an order of magnitude larger than the transfer width: $\overline{\Delta} > 10w = 2000$ Å. The specular peak height can be scanned through in- and anti-phase scattering also at constant $\vartheta_i = \vartheta_f$ by varying continuously the He wave length, i.e. the beam energy. This has been done by Tenner, Spruit, Kuipers, Kleyn [1987].

We have ignored in the discussion above that in analogy with isolated adsorbates and defects (see Chap. 4) also the step edges are expected to contribute to the He diffuse scattering. In the absence of this diffuse scattering and when the mosaic structure can be neglected, the peak height under in-phase conditions should be the same, independent of the step density. Figure 6.4 shows that this is obviously not the case. The peak heights of the He beam specularly reflected from a low step density surface (Fig. 6.4b) are consistently larger than peak height maxima obtained from the higher step density surface (Fig. 6.4a) under in-phase conditions. The mosaic distribution of Pt crystals being very narrow, the difference is mainly due to diffuse scattering around the step edges. The de-

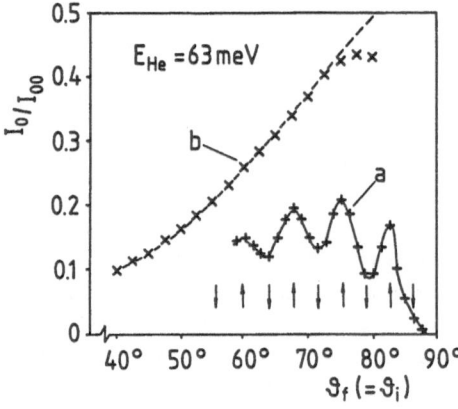

Fig. 6.4. Height of the specular He beam ($E_{He} = 63$ meV), relative to that of the incident beam, as a function of $\vartheta_f(= \vartheta_i)$. The data were obtained for Pt(111) surfaces after two different preparation procedures (see text). The arrows indicate in-phase ($\uparrow$) and anti-phase ($\downarrow$) scattering conditions for $E_{He} = 63$ meV corresponding to a Pt(111) interlayer spacing of 2.27 Å

termination of the size of the cross-section for diffuse scattering associated with steps will be discussed in the next section.

The TEAS procedure to characterize the structure of surfaces, as exemplified above for Pt(111), is by no means restricted to close packed (metal) surfaces. It can be applied in general. As an example we consider the case of an fcc (115)-surface being composed of a regular array of 2.5$a$ ($a$ being the nearest neighbour distance) wide (001)-terraces separated by monatomic (111)-steps. Obviously, also in this case the existence of mosaic structure and/or non-perfect preparation procedures will lead to (small) deviations from the desired surface structure. This is illustrated in Fig. 6.5 showing a gross deviation from the ideal periodic structure, due to the presence of two 1.5$a$ wide (001)-terraces instead of the regular width of 2.5$a$. In fact, these irregularities separate areas with a perfect (115)-structure, i.e. (115)-domains. Inspection of Fig. 6.5 shows that these domains appear at a different level; the level difference between adjacent domains amounts to $h_b = a\sqrt{2/27}$. Associated with these (115)-domains at different levels the (115)-crystal has "steps", which will subsequently be called domain boundaries in order to avoid confusion with the regular periodic (111)-steps. In general, the boundary height on a (11$m$)-face corresponds to

$$h_b = a\left(\frac{2}{2+m^2}\right)^{1/2} . \tag{6.7}$$

The boundary height will, according to (6.6), lead to path length differences between scattering contributions from adjacent domains. Consequently, the presence of domain boundaries will be revealed by an oscillatory behaviour of the He specular peak height versus the perpendicular momentum change (compare Fig. 6.4 and the related discussion). This has been observed indeed by Conrad et al. [1986a] for Ni(115) at $T_s = 100\,\text{K}$ as illustrated in Fig. 6.6. Note that the peak width is plotted rather than the peak height. The results, however, are equivalent since the width and the height of the specular peak behave in a complementary way as argued above. The experimental data were obtained by directing the incident beam perpendicular to the ledges in the step-up direction, i.e. along the $\overline{552}$-azimuth. The oscillations of the peak width (FWHM) are clearly visible; the maxima and minima are related with anti-($\downarrow$) and in-phase ($\uparrow$) scattering conditions, respectively, corresponding to the elementary boundary height of 0.68 Å obtained from (6.7). Assuming a geometric distribution of domain widths Conrad et al. [1986] deduce a mean domain width, i.e., average

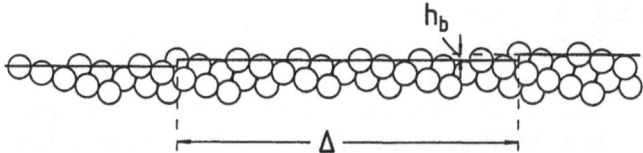

**Fig. 6.5.** Schematic view of a non-perfect fcc-(115) surface, $\alpha = 18.8°$ instead of $\alpha = 15.8°$, corresponding to an ideal (115) surface. The height difference $h_b$ between adjacent (115) domains (see text) and the distance $\Delta$ between domain boundaries are indicated

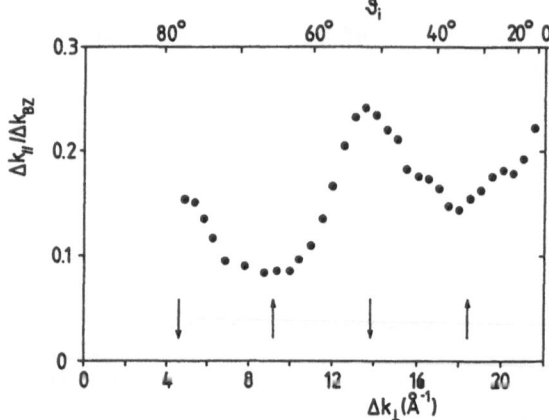

Fig.6.6. Experimental peak width versus perpendicular momentum change for 63 meV He scattered along the [$\overline{5}52$]-azimuth of a Ni(115) surface (after Conrad et al. [1986a]). The arrows indicate in-phase (↑) and anti-phase (↓) scattering corresponding to a height difference between adjacent domains given by (6.7)

flat (115) areas of $50 \pm 5$ Å. This value would correspond to a mean deviation of about 0.8° from the desired ideal (115)-orientation. The authors obtain similar results from experiments conducted along the [1$\overline{1}$0]-azimuth, i.e., parallel to the (111)-steps. Experiments on Cu(115) by the Saclay group [Fabre, Gorse, Lapujoulade, Salanon, 1987b] show a similar broadening of the anti-phase peaks.

We turn in the next section to low-index surfaces in order to analyze in detail the diffuse scattering from "random" steps.

## 6.4 The Cross-Section for Diffuse Scattering from Steps

Let us assume that the cross-section for diffuse He scattering from steps has the shape of a strip of width $D$ along the step edge contour. Denoting by $S$ the step edge length per unit surface area (a quantity directly related to the step density), a fraction $DS$ of the surface will scatter diffusely. Thus, in analogy with the scattering from adsorbates and monovacancies (3.4'), the in-phase fractional specularly reflected He amplitude is

$$A/A_0 = 1 - DS \tag{6.8}$$

and the relative He intensity is given by

$$I/I_0 = (1 - DS)^2 \tag{6.9}$$

with $I_0$ the specular intensity for $S \approx 0$.

In order to obtain a reliable value for the cross-section width $D$, an accurate determination of the step density (the step length $S$ per unit surface area) is needed. This has been done by the "facet ensemble" analysis of the peak shapes outlined in the preceding section and described in detail by Verheij, Lux, Poelsema [1984]. The randomly stepped surface has been prepared from a low step density ($\approx 0.001$) Pt(111) surface by the high temperature ion bombardment procedure described in Sect. 6.5.3. The specular He peak shape has been monitored very accurately under anti-phase conditions (Fig. 6.7). The analysis of this peak profile leads to the terrace width distribution shown in Fig. 6.8 and to an average terrace width of $\overline{\Delta} = 84.2\,\text{Å}$ (step atom density 0.0285), corresponding to $S = 1.2 \times 10^{-2}\,\text{Å}^{-1}$ if one assumes that the steps are oriented along $\langle 011 \rangle$ azimuthal directions. The solid curve in Fig. 6.7 represents the best fit corresponding to this step density. Note that the value of $\overline{\Delta}$ (not necessarily also the detailed terrace width distribution in Fig. 6.8) is independent of the model assumption the peak profile analysis is based upon [Verheij, Poelsema, Comsa, 1985].

The in-phase peak height is now measured on this same surface and on the surface with low step density ($\approx 0.001$). These values provide the ratio $I/I_0$ in (6.9); inserting in addition the step length per unit area determined above, $S = 1.2 \times 10^{-2}\,\text{Å}^{-1}$, a cross-section width $D = 12 \pm 2\,\text{Å}$ is obtained [Verheij, Poelsema, Comsa, 1985]. The value applies for $E_{\text{He}} = 16\,\text{meV}$ and $\vartheta_i = 52°$. Note that the size of $D$ compares well with the diameter of about $10\,\text{Å}$ of the cross-section for diffuse scattering from monovacancies ($\Sigma = 150\,\text{Å}^2$, see Sect. 4.4) under similar experimental conditions. Recently, a value of $D = 13\,\text{Å}$ has been

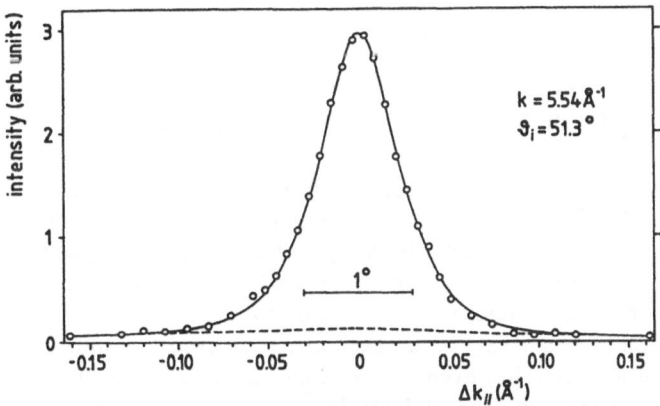

**Fig. 6.7.** Profile of the specular He peak ($E_{\text{He}} = 16\,\text{meV}$) measured under anti-phase conditions on a Pt(111) surface with a random step distribution (average terrace width $\overline{\Delta} \approx 84.2\,\text{Å}$) prepared by 600 eV Ar$^+$-ion bombardment at $T_s = 690\,\text{K}$ (fluence $10^{15}$ Ar$^+$ ions/cm$^2$)

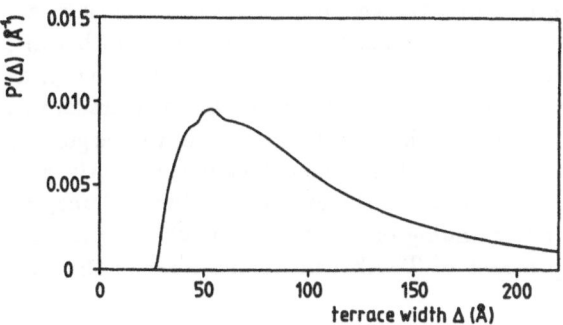

**Fig. 6.8.** Terrace width distribution deduced from the analysis of the specular He peak profile in Fig. 6.7 by applying the facet model (see text). Plotted is the probability of finding an atom on terraces of a certain width versus the terrace width

obtained for steps on a Cu(001)-surface [Sánchez, Ferrer, 1987]. Consequently, 10 Å can be regarded as a "general" length scale associated with diffuse He scattering.

Summarizing, we have presented three ways to infer the step atom density, and therefore the mean terrace width, from TEAS data. The most accurate but also most cumbersome way, i.e., the complete analysis of peak profiles taken under anti-phase conditions, supplies exact values for the step atom density. In many instances, however, one is satisfied with less accurate though faster estimates. This is provided either by measuring the reduction of the in-phase peak height with increasing step density when $D$ is known or by measuring the oscillatory behaviour of the He specular peak height as a function of perpendicular momentum change. The first one is also rather accurate. The application of this procedure, however, requires the ability to compare with a surface "without" steps, i.e. $S \approx 0$. In the other case the observation of an oscillatory behaviour of the He specular peak height with varying perpendicular momentum change (Fig. 6.4) enables a rough estimate of the step density without additional measurements. Note, however, that oscillations will be visible only if the mean terrace width is not much larger than the transfer width of the instrument and not smaller than the characteristic "10 Å size".

## 6.5 The Influence of Temperature on Surface Morphology During Ion Bombardment

The surface damage induced on a Pt(111) surface by ion bombardment at low temperatures has been discussed in detail in Sect. 4.4. In short, monovacancies created at $T_s = 80\,\mathrm{K}$ are thermally immobile. Only immediately subsequent to the sputtering event — and due to the energy deposited by the impinging ion — may the newly created monovacancies jump once or twice in the direction of an already existing vacancy: vacancies attract each other. It has been shown, in addition, that even if in a sputtering event more than one vacancy is created, these are always monovacancies created at least 10 Å apart. As a consequence of this behaviour, the distribution of monovacancies is largely random up to densities of about 0.04; at higher densities a restricted vacancy clustering is observed [Poelsema, Comsa, 1985b; Comsa, Poelsema, 1985; Poelsema et al., 1985].

High temperature annealing is a well-known procedure to heal surface defects, in particular those created by ion bombardment. The vacancies become mobile and, due to their mutual attraction, cluster into islands. Obviously, as already mentioned, an island of vacancies represents in fact a new terrace, i.e., an almost defect free surface area. The higher the temperature, the larger the vacancy mobility and correspondingly the island size. Thus at sufficiently high temperatures, the shape of the surface prior to the ion bombardment damaging is recovered. A more detailed understanding of this process is also of practical importance. The access to it by means of TEAS will be discussed in the next three sections.

We will take advantage of the fact that thermal He scattering and ion bombardment do not influence each other; i.e., we can monitor the surface morphology by TEAS *during* ion bombardment at various surface temperatures. We can thus follow in real time the interplay between defect creation and annealing. The measurements are done in both the in- and the anti-phase mode. As shown in the preceding section, the in-phase specular peak height is a measure of the defect density (vacancies and step atoms). As long as no interference between He atoms scattered from terraces separated by monatomic steps takes place (initially step-free surface and "vacancy islands" less than $\approx 12$ Å in diameter) the in- and the anti-phase specular peak heights are indistinguishable. However, when the vacancy islands, i.e. the new terraces, become larger than $\approx 12$ Å — the width of the diffusely scattering step edge strip — they start to reflect specularly also. Due to the interference effects dicussed above the anti-phase peak becomes broader and the anti-phase peak height smaller than the in-phase one. New information on the detailed morphology of the surface becomes accessible.

### 6.5.1 In-phase Peak Height During Ion Bombardment at Various Temperatures

The in-phase He specular peak height $I$ has been monitored during constant flux $(1.5 \times 10^{13}$ ions/cm$^2$s) of a 600 eV Ar$^+$ bombardment at various fixed Pt(111)-surface temperatures. In Fig. 6.9 the normalized specular peak height $I/I_0$ is plotted as a function of the number of sputtered monolayers. The normalization to $I_0$, the initial peak height (i.e. defect free surface) automatically corrects for Debye–Waller effects. The measured fluence (i.e. flux times bombardment time) is transformed into sputtered monolayers by using the sputtering yield of 600 eV Ar$^+$ determined on the same surface by the method described below in Sect. 6.5.3. Since the sputtering yield is not expected to depend significantly on the surface temperature [Roosendaal, 1981], the dramatic influence of the surface

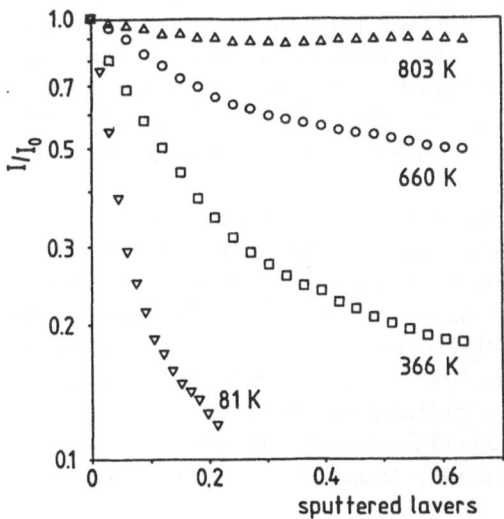

**Fig. 6.9.** In-phase ($E_{He} = 16$ meV, $\vartheta_i = 42°$) specular He peak height during sputtering with 600 eV Ar$^+$ ions as a function of the number of sputtered Pt (111) monolayers at various surface temperatures. The peak heights are normalized at their initial ("perfect" surface) values

temperature on the shape of the damaging curves as is obvious in Fig. 6.9 has to be attributed to the strong increase of the annealing rate with temperature [Poelsema, Verheij, Comsa, 1984a].

The $T_s = 80$ K curve is identical to that in Fig. 4.3. As discussed already, in this temperature range monovacancies are thermally immobile. No annealing takes place up to about 180 K; the damaging curves appear to be identical in the whole range $80 - 180$ K. Above 180 K, the shape of the damaging curves becomes notably temperature dependent: their slope decreases gradually with increasing surface temperature. At a given $Ar^+$-ion fluence the density of defects (vacancies and step atoms) decreases with increasing temperature. The defects anneal faster at higher temperatures because of the higher vacancy mobility. At $T_s = 803$ K the reflectivity remains almost constant after a small initial decay. This indicates that at high temperatures already at relatively low ion fluences the rates of defect creation and annealing become equal: the ledge concentration reaches rapidly a steady state value.

### 6.5.2 Anti-phase Peak Height During Ion Bombardment at Medium Surface Temperatures

As discussed in the introductory part of Sect. 6.5 above, the in- and anti-phase peak heights are indistinguishable as long as no new terraces with a size larger than the diffuse scattering step edge strip ($\approx 12$ Å) develop. Thus below 180 K, where the vacancies are thermally immobile, the in- and anti-phase damaging curves are obviously identical. But also above this temperature, say around 300 K — where the vacancies are already mobile and a certain amount of annealing takes place (see Fig. 6.9) — the two damaging curves still stay practically identical as shown in Fig. 6.10. This means that the vacancy clustering is limited to islands smaller than $\approx 12$ Å. It is only at higher temperatures that the size of vacancy islands (the "new" terraces) becomes substantially larger. This is demonstrated by the damaging curves taken at 500 K and shown in Fig. 6.11. While the in-phase peak height is roughly halved upon an $Ar^+$ fluence of $1.5 \times 10^{15}$ cm$^{-2}$ ($\approx 3$ sputtered monolayers) the anti-phase peak height falls nearly exponentially, being attenuated by a factor of 25 at a three times smaller fluence. This shows directly, that on the one hand the linear dimensions of the new terraces substantially exceed 12 Å and on the other hand they stay well below the transfer width of the instrument $w = 200$ Å. (Otherwise, the anti-phase peak height would not be reduced so dramatically.)

The nearly exponential decay of the anti-phase peak height (Figs. 6.11 and 6.12) can be rationalized by the formation of three-dimensional defect structures. (The formation of such structures has already been demonstrated by Verheij, Van den Berg, Armour [1982] during ion bombardment of Ni(100) in a similar temperature regime; see also Cherns [1977] and Verheij [1982].) It is remarkable that the damaging curves can be quantitatively described by making only two reasonable assumptions: (1) atoms are sputtered only from uncovered layers and (2) the vacancy mobility (leading to island formation) takes place only within a given layer, but not between layers (no mass transport between layers).

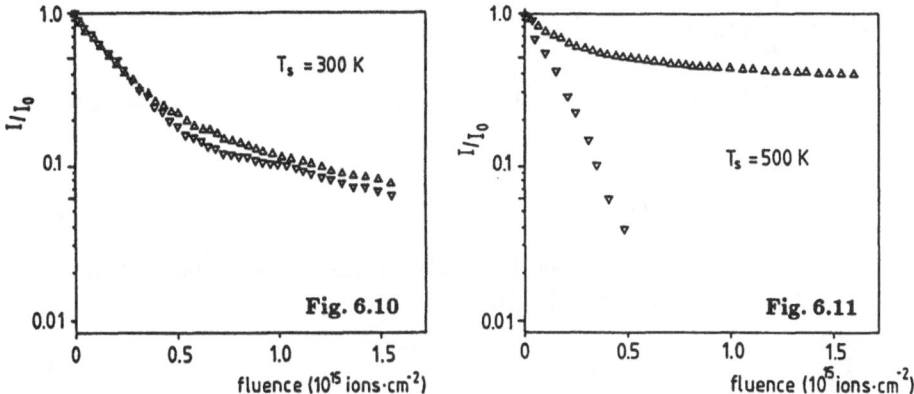

**Fig. 6.10.** In-phase ($\triangle$; $\vartheta_i = 41.4°$), and anti-phase ($\triangledown$; $\vartheta_i = 46.3°$) specular He peak height ($E_{He} = 63\,\text{meV}$) during sputtering with $600\,\text{eV}$ $Ar^+$ ions versus the ion fluence on a Pt(111) surface at $T_s = 300\,\text{K}$. The He peak heights are normalized at their respective initial "perfect" surface values

**Fig. 6.11.** As in Fig. 6.10, but at $T_s = 500\,\text{K}$

Let us denote by $Y$ the sputtering yield and by $x_n$ the uncovered fractional area of layer $n$ (the layers are numbered starting with the outermost layer $n = 0$). The number of atoms sputtered from the $n$th layer by $dF$ impinging ions equals thus $dN_n = x_n Y\,dF$; the corresponding sputtered area is $dN_n/n_s$ where $n_s$ is the atom density in each layer ($n_s = 1.5 \times 10^{15}\,\text{atoms/cm}^2$ for the Pt(111) surfaces). Accordingly, by sputtering the atoms of the $n$th layer its fractional area decreases by $dx_n = -dN_n/n_s$ ($dF$ — the differential fluence — represents ions incident on unit area). Simultaneously, of course, a fractional area $dx_{n-1} = -dN_{n-1}/n_s = -x_{n-1}Y\,dF/n_s$ of the $(n-1)$th layer is also sputtered away, uncovering a corresponding area of the $n$th layer. Thus, the total change of the fractional area of the $n$th layer is:

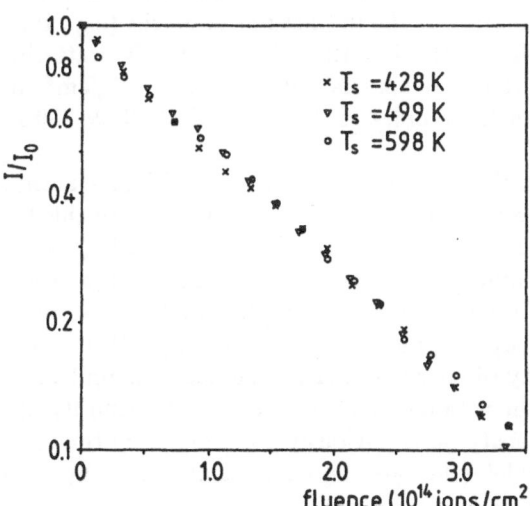

**Fig. 6.12.** Anti-phase ($E_{He} = 16\,\text{meV}$, $\vartheta_i = 51.5°$) specular He peak height during sputtering with $600\,\text{eV}$ $Ar^+$ ions vs ion fluence at various Pt(111)-surface temperatures: $T_s = 428\,\text{K}$ ($\times$), $T_s = 499\,\text{K}$ ($\triangledown$) and $T_s = 598\,\text{K}$ (o). The data are normalized at their respective initial "perfect" surface values

65

$$dx_n = \frac{Y}{n_s}(x_{n-1} - x_n)dF \quad .$$

Taken for all layers, this represents a system of coupled differential equations. The solution is given by

$$x_n = \frac{\left(\frac{Y}{n_s}F\right)^n}{n!}\, e^{(-Y/n_s)F} \quad . \tag{6.10}$$

For anti-phase conditions the resulting relative amplitude of the He waves scattered in the specular direction is given by

$$\frac{A}{A_0} = x_0 - x_1 + x_2 - x_3 + \ldots = e^{(-2Y/n_s)F} \quad , \tag{6.11}$$

as long as the width of the created terraces is small compared to the transfer width. The relative intensity is then given by

$$\frac{I}{I_0} = e^{(-4Y/n_s)F} \quad . \tag{6.12}$$

Thus this simple model predicts indeed the exponential decay of the antiphase peak height with increasing fluence. A particularly interesting feature of (6.12) is that the slope of $\ln I/I_0$ is independent of surface temperature (i.e. also of the size of the newly created terraces). This unexpected prediction of the model is remarkably confirmed by the experiment in a relatively large temperature range $400\,\mathrm{K} < T_s < 600\,\mathrm{K}$ (see Fig. 6.12). The three plotted curves coincide without the use of any free parameter. Retrospectively, this confirms also the beautiful prediction of (6.10), that the uncovered fractional areas of the various layers in the three dimensional structures, $x_n$, have a Poisson distribution.

The common slope $(-4Y/n_s)$ of the curves in Fig. 6.12 supplies directly the value of the sputtering yield, $Y = 2.5$ atoms/ion for $600\,\mathrm{eV}$ $Ar^+/Pt(111)$. This value will be confirmed by a completely different approach in Sect. 6.5.3 below.

The slope in Fig. 6.12 being independent of the terrace width, the step density has to be obtained from the attenuation of the in-phase peak height (e.g., from plots like those in Fig. 6.9) and the width $D = 12\,\text{Å}$ of the diffusely scattering strip along the step edges. For instance, at a fluence of $1.5 \times 10^{15}$ $Ar^+/cm^2$ the average terrace width at $T_s = 500\,\mathrm{K}$ becomes $\overline{\Delta} \simeq 60\,\text{Å}$ and at $T_s = 780\,\mathrm{K}$: $\overline{\Delta} \approx 500\,\text{Å}$.

Thus, as intuitively expected, the size of the vacancy islands (2D-vacancy clusters) increases with the surface temperature. One way to rationalize this behaviour is the following. Upon creation the monovacancies are migrating. When they meet, they grow into larger and larger clusters. Let us assume that when these vacancy clusters reach a critical size ($n_c$-vacancies), they become immobile. Their size continues to grow by merging of vacancy clusters with $n < n_c$, which are still mobile. The density of cluster-nuclei $\varrho_\nu$ (number per unit area) determines ultimately the size of the vacancy islands: the smaller the density, the larger the islands. In turn, the density $\varrho_\nu$ is obviously determined by the critical size $n_c$ and by the "length scale" (the ratio between vacancy mobility and vacancy creation rate). Thus an increase in surface temperature decreases the nuclei density $\varrho_\nu$ via the increased vacancy mobility in two ways: (a) the critical size $n_c$ becomes larger (the size of the cluster has to be larger in order to become immobile) and (b) the increased "length scale" favours also larger islands (the under-critical clusters being more mobile may reach already immobile supercritical ones before becoming supercritical themselves by coalescence with newly created vacancies).

For simplicity one may introduce a characteristic length $L$, the vacancy islands size, which is obviously proportional to $\varrho_\nu^{-1/2}$. This length $L$ increases with surface temperature and decreases with the vacancy creation rate (the product of the ion flux and the sputtering yield, $Y$). Note that the characteristic length $L$ is also a measure of the average path length of a vacancy from its creation to its encounter with an immobile cluster.

### 6.5.3  Anti-phase Peak Height at Higher Temperatures

The behaviour of the anti-phase peak height as a function of ion fluence changes dramatically above about 700 K: instead of decreasing exponentially (Fig. 6.12) the peak height starts to oscillate (Fig. 6.13). The oscillations are damped leading eventually to a stationary value of $(I^{\text{anti}}/I^{\text{in}})_{\text{stat}}$. At this stage the average terrace widths can be inferred by one of the methods described above. For the curves in Fig. 6.13 one obtains: $\overline{\Delta}_{747\,\text{K}} \approx 320\,\text{Å}$, $\overline{\Delta}_{783\,\text{K}} \approx 500\,\text{Å}$ and $\overline{\Delta}_{803\,\text{K}} = 1000\,\text{Å}$. Thus at least in the stationary stage the picture of an average terrace width and thus of a characteristic length $L$ increasing with increasing temperature appears to continue to be valid also in the high temperature range.

Note that in contrast to the oscillations in Fig. 6.4a obtained by varying the scattering conditions ($\vartheta_i$ or $E_{\text{He}}$), the oscillatory behaviour in Fig. 6.13 appears as a function of ion bombarding time, keeping *all* experimental conditions fixed.

The appearance of oscillations indicates that in the high temperature range the defect annealing process is qualitatively different from that at medium temperatures. As will be shown below the simultaneous sputtering and annealing

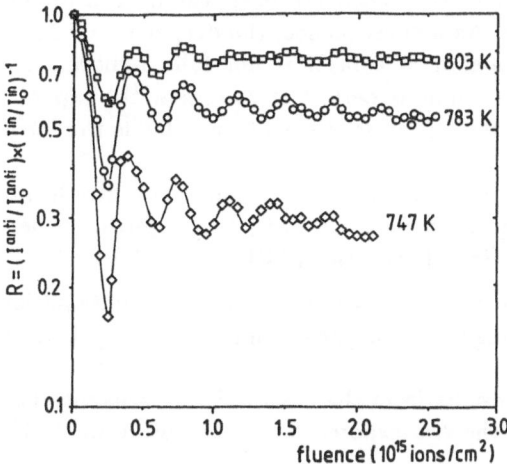

Fig. 6.13. Ratio of the normalized anti-phase ($\vartheta_i = 51.5°$) He specular peak heights $(I/I_0)^{\text{anti}}$ to the in-phase ($\vartheta_i = 41.6°$) ones $(I/I_0)^{\text{in}}$ during sputtering with 600 eV Ar$^+$ ions vs ion fluence. The data were taken with a $E_{\text{He}} = 16\,\text{meV}$ He beam at various surface temperatures: $T_s = 803\,\text{K}$ (□), $T_s = 783\,\text{K}$ (o) and $T_s = 747\,\text{K}$ (◇). In order to emphasize the role of interference effects, the data are corrected for diffuse scattering (and also for Debye–Waller effects) by plotting the ratio $(I/I_0)^{\text{anti}}$ to $(I/I_0)^{\text{in}}$ on the ordinate

lead no longer to the build-up of three dimensional defect structures but can be viewed as a "layer-by-layer" removal of the (111) layers of the Pt(111) surface.

Let us discuss first the limiting case of very high temperatures (very large vacancy mobility, i.e. $L \rightarrow \infty$) and assume that, when the sputtering starts, the surface is a unique, defect-free (111) terrace (only the layer $n = 0$ is uncovered). Due to their large mobility all the vacancies which start to be created in this layer coalesce in a unique island which represents the newly growing terrace $n = 1$. This lower lying terrace $n = 1$ is separated from the original terrace $n = 0$ by a monatomic ledge (step-up). Continuing the ion bombardment, vacancies are now created in both layers; i.e., in the not yet removed area of layer $n = 0$ and in the uncovered area of layer $n = 1$. The vacancies in layer $n = 0$ move around until they reach the ledge, and add up to the vacancy island: the new terrace $(n = 1)$ grows at the expense of the old, upper one $(n = 0)$. The monatomic ledge is a sink for vacancies in the upper layer $n = 0$. However, with the assumption (2) made in the case of the medium temperature range (Sect. 6.5.2), i.e., "no mass transport between layers", the vacancies in the lower layer, $n = 1$, will not "disappear" when reaching the ledge (layer $n = 0$ atoms are not allowed to fill up layer $n = 1$ vacancies). Thus the lower layer $n = 1$ vacancies will accumulate and eventually coalesce in a new vacancy uncovering the next deeper layer $n = 2$. This is in fact the microscopic mechanism of the creation of three dimensional defect structures at medium temperatures discussed in the preceding section. It appears now obvious that in order to get a "layer-by-layer" removal, we have to give up the "no mass transport" between layers assumption. We may assume that the filling of lower layer vacancies close to the "step-up" ledge with atoms of the upper layer is an activated process, which is hindered at medium temperatures ("no mass transport between layers") but becomes easily possible at the high temperatures considered here. Thus the ledge becomes also a sink for vacancies created in the lower layer, which — due to their large mobility at high temperatures — do not have the opportunity to coalesce and thus to uncover the deeper $(n = 2)$ layer.

In this picture all vacancies, whether created in the upper $(n = 0)$ or in the lower $(n = 1)$ layer, end at the ledge separating the two layers; irrespective of the layer in which they are created, the vacancies disappear by consuming exclusively atoms of the *upper* layer. As a consequence, the deeper $n = 2$ layer is not uncovered as long as the upper $n = 0$ layer is not fully removed. At this moment, the initial situation — a unique perfect terrace, now the exposed layer $n = 1$ — is restored and the whole process starts again. This is the ideal "layer-by-layer" removal.

Let us now estimate the variation of the in- and anti-phase specular beam peak height vs ion fluence during such an *ideal* "layer-by-layer" process when monitored with an *ideal* instrument (infinite transfer width, $w = \infty$):

i) The in-phase peak height will stay essentially constant, keeping its initial value $I_0^{in}$ because the diffuse scattering from the single ledge, separating the two terraces, is certainly negligible.

ii) The anti-phase peak height results from the destructive interference between the He waves scattered from the two exposed layers. Denoting by $f$ the sputtered amount in units of monolayers (a quantity proportional to the fluence), for $f \leq 1$ the remaining fractional area of the upper layer is $1 - f$ and correspondingly, the uncovered area of the lower layer $f$. Thus the relative amplitude scattered in the specular direction will be

$$(A(f)/A_0)^{\text{anti}} = (1 - f) - f = 1 - 2f \qquad (6.13)$$

and the relative peak height:

$$(I(f)/I_0)^{\text{anti}} = (1 - 2f)^2 \quad . \qquad (6.14)$$

After having sputtered a full monolayer, $f = 1$, the initial situation is — as we have seen above — fully restored and (6.14) gives correspondingly $I^{\text{anti}}(f = 1) = I_0^{\text{anti}}$. By continuing the sputtering the curve segment described by (6.14) is repeated for each new sputtered monolayer by inserting $f = f' - n$, with $f'$ the whole sputtered amount and $n$ the number of full monolayers sputtered ($0 \leq f \leq 1$). The curve obtained is shown in Fig. 6.14; it exhibits an oscillatory, non-damped behaviour. Note that each time when half a monolayer is sputtered away, i.e. $f = 0.5$, the anti-phase specular peak height passes through a minimum $I^{\text{anti}}(f = 0.5) = 0$; at this point the areas of the remaining upper and of the uncovered lower layer are equal and the destructive interference is complete.

However, such a doubly idealized experiment can not be realized; not even in a first approximation. This is primarily due to the fact that any real instrument has a finite transfer width, $w$ (here $w \approx 200$ Å). The anti-phase peak height $I^{\text{anti}}$ vs fluence measured with such an instrument on a surface, where the process takes place under the ideal conditions defined above, will show no oscillation whatsoever. The characteristic length $L$, and thus the average terrace width $\overline{\Delta}$ being infinite, the ratio $\overline{\Delta}/w$ becomes also infinite. The instrument "sees" within the size of the transfer width over the almost entire surface only one terrace (either the upper or the lower one); no interference and thus no anti-phase peak broadening is measured.

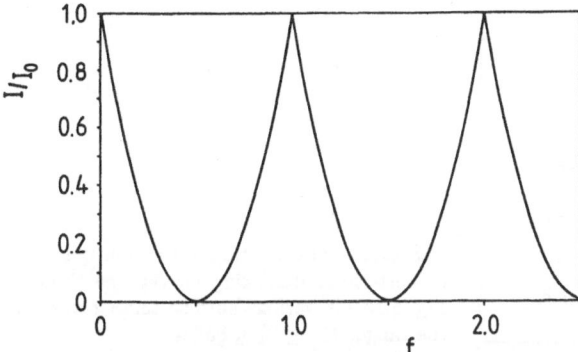

**Fig. 6.14.** Expected relative He specular peak height as a function of the number of sputtered layers, $f$, in the case of ideal layer-by-layer removal measured with an ideal instrument (infinite transfer width) in anti-phase mode; see text

In order to see oscillations the characteristic length $L$ (and $\overline{\Delta}$) has to be reduced to values below $\sim 10\,w$ by reducing the temperature of the surface. Because the sputtering process proceeds now under "non-ideal" conditions ($L$ and $\overline{\Delta}$ finite) the vacancies will form a number of vacancy islands (no longer a unique one) distributed randomly on the surface. By fortuitous island coalescence, the dimensions of some of the islands (lower level terraces $n = 1$) will exceed the characteristic length $L$. Thus new vacancy islands may be formed on these large lower level terraces ($n = 1$) uncovering the next lower terrace ($n = 2$), *before* the upper layer ($n = 0$) is completely removed in spite of the fact that interlayer mass transport is still taking place. The process is no more an ideal "layer-by-layer" sputtering, but gets gradually out of phase. The too early appearance of vacancy islands in the next lower layer may be additionally enhanced by the fact that when the temperature is reduced, the probability of the mass transport between layers (filling of lower layer vacancies by upper layer atoms) might be reduced also. The desynchronization (on a scale large with respect to $w$) of the removal process along the surface is the cause of the damping of the oscillations observed experimentally (Fig. 6.13). Note, however, that at least during the first half period, i.e. until the first half monolayer is removed, the process proceeds synchronously along the whole probed area ($\approx 1\,\mathrm{mm}^2$). Indeed, the positions of the first minima measured at different temperatures coincide within experimental accuracy: $2.5 \times 10^{14}$ "$600\,\mathrm{eV}$" $\mathrm{Ar}^+/\mathrm{cm}^2$. This is apparent in Fig. 6.15, showing that the fluence corresponding to the first minimum does not move in spite of varying the characteristic length $L$ by about one order of magnitude ($200\,\text{Å} < \overline{\Delta} < 2000\,\text{Å}$ for $713\,\mathrm{K} \leq T_s \leq 856\,\mathrm{K}$). The fact that the peak-height minimum value is not zero, is not a manifestation of an asynchronism, but simply results from the ratio $\overline{\Delta}/w$ being of the order of/or larger than unity ($0.7 < \overline{\Delta}/w < 6.4$ in Fig. 6.15).

Summarizing the discussion in this section, we note that the oscillations of the anti-phase peak height observed during ion bombardment at high temperatures are the first evidence for "layer-by-layer" sputtering of crystals. Besides this, two novel applications of TEAS emerge which might become important in future research:

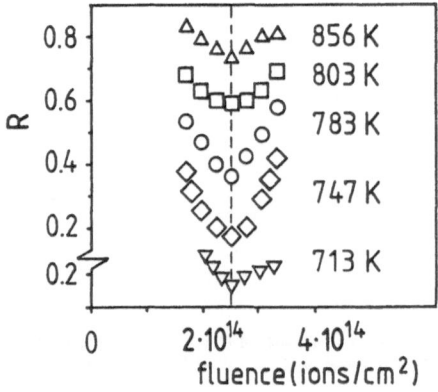

Fig. 6.15. The location of the first minimum of anti-phase damaging curves like those in Fig. 6.13 for various surface temperatures in the range: $713 \leq T_s \leq 856\,\mathrm{K}$

1) *Interferometric Measurement of Sputtering Yields.* The position of the first minimum in the oscillatory curve gives directly the number of incident ions necessary to sputter half a monolayer. This is probably the most simple and accurate method to determine sputtering yields; an additional advantage is that during the first half period the surface is still of excellent quality, i.e. the measured yield is not influenced by an unknown surface morphology. The result for 600 eV $Ar^+$ ions on Pt(111), $Y \approx 3$ atoms/ion, agrees well with the value derived from the data obtained at "medium" temperatures (see Sect. 6.5.2, Fig. 6.12).

2) *In Situ Preparation of Surfaces with Given Step Density.* A desired step density ($\approx 1/\Delta$) can be obtained by sputtering with a fixed ion flux and by choosing the right temperature. The stationary structure, obtained when the amplitude of the oscillation has vanished, can be frozen by simultaneously interrupting the ion beam and rapidly cooling the crystal.

The efficiency of freezing the surface structure has been tested by comparing the step densities before and after the freezing procedure. Detailed studies have also shown that the step structures produced in this way are preferentially oriented along the high symmetry lines of the surface [Lahee, Manson, Toennies, Wöll, 1986]. Starting with an almost defect free surface, this procedure thus allows to create in situ surfaces with any desired defect density in the range $0.001 < \Theta_d < 0.15$. This is very helpful when investigating the role of steps in surface processes; e.g. so far, in order to measure the reactivity as a function of step density, a number of crystals cut at various angles $\alpha_i$ had to be prepared, monitored, cleaned and measured sequentially in the UHV apparatus. With the proposed method (already applied for the system $H_2/Pt$, see Chap. 8) all measurements at various step densities could be performed on the *same* Pt crystal (i.e. with the same amount of intrinsic impurities) in a unique run, *without* breaking the vacuum. This is particularly important for the accuracy and reproducibility of the results.

## 6.6 Investigation of Crystal Growth

The discussion above on sputtering can be applied without restriction to crystal growth: sputtering being the inverse of crystal growth. This is exemplified for the case of the evaporation of Cu from the vapour phase on Cu(001) [Gomez, Bourgeal, Ibañez, Salmerón, 1985; De Miguel et al., 1987] in Fig. 6.16. The results are quite analogous to those for high temperature ion damaging of Pt(111) shown in Sect. 6.5.3. Using the same arguments we infer that the observed oscillations demonstrate layer-by-layer crystal growth.

The crystal growth case being of more general interest than crystal damaging, it has received considerably more attention sofar. Reflection High-Energy Electron Diffraction (RHEED) is mostly applied to study crystal growth. RHEED intensity oscillations during growth [Van Hove et al., 1983; Neave, Joyce, Dobson, Norton, 1983], similar to those in TEAS [Gomez, Bourgeal, Ibañez, Salmerón,

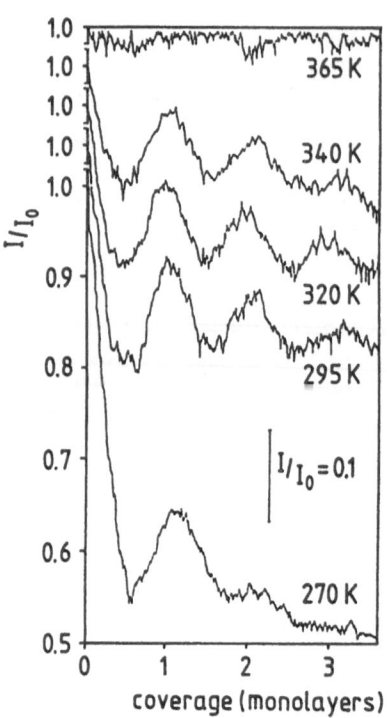

Fig. 6.16. Normalized anti-phase ($E_{He} = 63$ meV, $\vartheta_i = 76°$) He specular peak heights during Cu deposition on a Cu(001) substrate, versus the deposited Cu amounts (in ML) (after Gómez, Bourgeal, Ibañez, Salméron [1985])

1985; De Miguel et al., 1987], have been observed. The origin of the RHEED oscillations is still a matter of considerable debate. There is no doubt that multiple scattering is important in RHEED [Joyce, Dobson, Neave, Zhang, 1986] and that absolute intensities can only be derived using dynamical theory [Kawamura, Maksym, 1985]. As a consequence, a unique interpretation of the oscillations in RHEED is still lacking. Some authors even favour an interpretation in terms of multiple scattering effects from a varying step density [Dobson, Joyce, Neave, Zhang, 1987]. In that case the observation of oscillations does not necessarily indicate a perfect "layer-by-layer" growth mechanism. Others interpret the oscillations of RHEED intensities as due to diffraction from terraces applying kinematical theory [Lent, Cohen, 1984]. Although recent X-ray scattering data for Ge/Ge(111) seem to support the latter approach, it was stressed also that important differences with RHEED do occur [Vlieg et al., 1988].

TEAS measurements may in fact contribute to the solution of this problem. The combination of in-phase data, supplying directly the step edge concentration, with anti-phase data, probing the terrace interferences, will be very useful for a detailed analysis of surface morphology during growth. Note that the absence of oscillations in the anti-phase data, combined with little or no decay of the in-phase He peak height indicates an almost ideal "layer-by-layer" growth with large terraces (see the discussion of the doubly idealized ion damaging monitoring in Sect. 6.5.3).

# 7. TEAS as a Probe of Thermal Roughness on Monocrystalline Surfaces

## 7.1 Introduction

The concept of a roughening transition on monocrystalline surfaces has been introduced theoretically by Burton, Cabrera and Frank [1951] (see also Weeks [1980]). It plays a crucial role in the understanding of the crystal growth and of the equilibrium shape of crystal surfaces. On an atomic scale the roughening of a clean, originally smooth surface is characterized by the vanishing of the free energy associated with the creation of steps. In other words, above the roughening transition the proliferation of steps is not restricted. This leads to a logarithmic divergence of the height-height correlation function with increasing distance parallel to the surface [Kosterlitz, Thouless, 1973; Kosterlitz, 1974]. Experimentally, the existence of a roughening transition on low-index surfaces has been reported for the He solid — superfluid interface, see e.g. [Avron et al., 1980]. The basic question whether the roughening transition on a low index metal surface occurs below the crystal melting temperature or not is still disputed [Mochrie, 1987; Zeppenfeld, Kern, David, Comsa, 1989].

Recently, a definition of roughening confined to the case of nominally stepped, i.e. high Miller index surfaces has been formulated: At the roughening temperature the free energy for creating a kink within a step vanishes [Villain, Grempel, Lapujoulade, 1985]. Thus at the roughening temperature, the already present smooth step rows meander strongly due to the proliferation of kinks. Since the free energy required to create a kink atom in a smooth ledge is lower than that needed to generate a ledge atom on a smooth surface, roughening is expected to occur at lower temperatures on nominally stepped surfaces than on low-index ones. Indeed, using TEAS and X-ray scattering the nominally stepped Ni and Cu surfaces have been shown to undergo a roughening transition well below the melting temperature [Den Nijs, Riedel, Conrad, Engel, 1985; 1986; Conrad et al., 1986a; 1986b; Lapujoulade, 1986; Fabre, Gorse, Salanon, Lapujoulade, 1987a; 1987b; Conrad, Alley, Blanchard, Engel, 1987; Liang et al., 1987; Fabre, Salanon, Lapujoulade, 1988a; 1988b].

It is the aim of the present chapter to show that TEAS is particularly suited to investigate surface roughening. The emphasis will be on the TEAS methodology rather than on the detailed physics of the roughening transition. We first discuss how surface roughening affects the He peak profile and subsequently show how the presence or absence of roughening can be concluded from measuring the He peak width or height as a function of surface temperature, from analyzing the complete He peak profile at high temperatures and from measuring diffuse elastic scattering at large He parallel momentum change with increasing surface temperature. Finally, the anomalous dependence of coherent He scattering from Cu(011) on the temperature is discussed in terms of anharmonicity of surface thermal vibrational amplitudes.

## 7.2 Surface Structure at Higher Temperatures

In order to investigate the temperature dependence of the surface structure Lapujoulade and coworkers have measured the intensity of He diffraction peaks from a variety of nominally stepped and low-index copper surfaces as a function of temperature [Lapujoulade, Perreau, Kara, 1983]. A selected survey of their results is given in Fig. 7.1; their data are corrected for Debye–Waller effects. The experimental data show a substantial drop with temperature above $400 - 1000$ K, depending on the surface orientation. Already in 1983 Lapujoulade, Perreau, and Kara suggested that this feature can be attributed to "some kind of thermal roughening of these surfaces".

The anomalous decay of the He peak intensities with increasing temperature has been investigated theoretically by Villain, Grempel and Lapujoulade (VGL) [1985] for the case of nominally stepped surfaces. They suggested that above the roughening transition temperature $T_R$ the steps become increasingly fuzzy, i.e. they meander, due to a vanishing free energy for the creation of kinks in the originally smooth ledges. VGL developed a model in terms of two parameters: a kink formation enegy $W_0$ and a step-step repulsive interaction energy $w$. A schematic representation of a (11m) surface below (a) and above (b) $T_R$ is given in Fig. 7.2. The creation of kinks on a regularly stepped surface leads to a broadening of the step-step distance distribution, while the step-step repulsion tends to maintain the original periodicity on a local scale. The combination of these — to some extent counteracting — effects leads to the formation of (11m) domains; as shown in Sect. 6.3 the average level of a domain differs from that of the adjacent ones by a height difference given by (6.7).

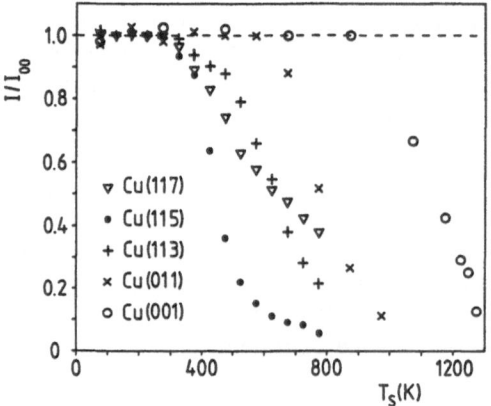

Fig. 7.1. Selected survey of He peak heights, normalized at their extrapolated $T_s = 0$ K values, versus surface temperature for various Cu monocrystalline surfaces. The data are plotted after correction for Debye–Waller effects [Lapujoulade, Perreau, Kara, 1983]

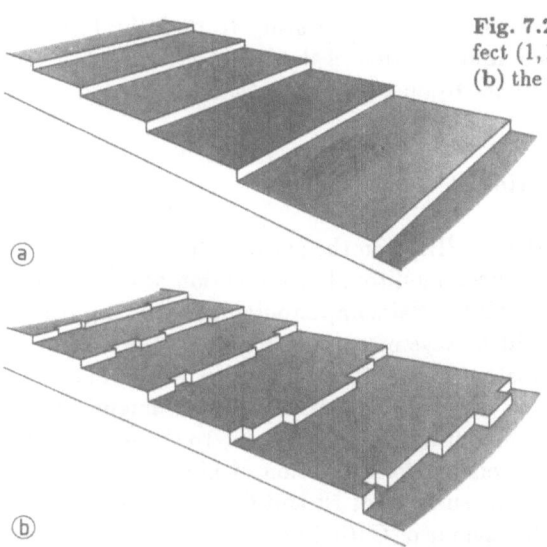

Fig. 7.2. Schematic representation of a perfect $(1, 1, 2n + 1)$ surface below (a) and above (b) the roughening transition temperature $T_R$

(a)

(b)

As discussed in detail in Chap. 6 the presence of domains separated by boundaries in a random up/down sequence can be revealed by TEAS through the coherent peak profile behaviour. The various ways to do this (in- vs anti-phase peak height behaviour, anti-phase peak width broadening or detailed peak profile analysis) are only various degrees of sophistication in the analysis of the same phenomenon. Den Nijs, Riedel, Conrad, Engel [1985; 1986] have been the first to recognize that the appearance of thermal roughening on nominally stepped surfaces can be revealed by measuring the in- and anti-phase peak widths as a function of temperature.

Note that the peak profile vs surface-temperature behaviour is both a general and a selective criterion for surface roughening. It is general because it is applicable to both low indexed surfaces, where the proliferation of steps leads to randomly stepped surfaces with a very characteristic peak-profile behaviour and, as shown by Conrad et al. [1986a; 1986b; Conrad, Allen, Blanchard, Engel, 1987] (see also next sections), also to nominally stepped surfaces. It is selective, because the characteristic behaviour of the peak profile discussed in the preceding chapter unequivocally demonstrates the appearance of surface domains separated by quantified height differences. In contrast, the simpler coherent peak intensity decrease vs temperature criterion, often used sofar, is by no means selective. Such a decrease can certainly be caused by roughening but — as will be shown below — can be observed also in the absence of roughening, being caused, e.g., by anharmonic effects [Jayanthi, Armand, Manson, 1985].

Thus, in short: the coherent peak-intensity behaviour may be an indication but is not a proof for surface roughening onset; for an unambiguous proof the behaviour of the peak profile has to be analyzed. This will be illustrated in the next sections by measurements on (11m)-nominally stepped surfaces, which do roughen, and on low-indexed surfaces, which apparently do not roughen in the investigated temperature range. While the peak intensity from Cu(011) has an anomalous temperature behaviour similar to that of the (11m) surfaces (Fig. 7.1), the peak profile behaviour will appear to be quite different.

### 7.2.1 The He Peak Width and Height

The in- and anti-phase peak widths (FWHM) on the (115) surfaces of Ni and Cu (Fig. 7.3a and b, respectively) have been monitored as a function of temperature by Conrad et al. [1986a] and Fabre, Gorse, Salanon, Lapujoulade [1987b], respectively. While the in-phase peak width stays essentially constant, the anti-phase peak width starts to increase substantially with temperature above 500 K for Ni(115) and above about 300 K for Cu(115). This evidences unambiguously the formation of (115) domains with different average heights due to the proliferation of kinks in the step-rows. The free energy for formation of kinks becomes zero around 500 K and 300 K for the (115) surfaces of Ni and Cu, respectively.

In contrast, the peak-profile behaviour of Cu(011) shows that no step proliferation, i.e. no roughening, appears up to 900 K, in spite of the fact that from the coherent peak-intensity decrease — like that in Fig. 7.1 but measured using X-ray diffraction — a roughening temperature of 850 K for Cu(011) has been inferred [Mochrie, 1987]. In Fig. 7.4 specular peak profiles monitored under-out-of phase conditions by Zeppenfeld, Kern, David, Comsa [1989] on Cu(011) by means of a high resolution He scattering instrument are shown. No peak broadening is seen up to the highest measured temperature (900 K), i.e. no (011)-terraces at various heights are produced.

The behaviour of the He-peak height vs surface temperature is consistent with that of the peak width. This is illustrated in Fig. 7.5a and b for Ni(113) [Conrad, Allen, Blanchard, Engel, 1987] and Cu(115) [Fabre, Gorse, Salanon,

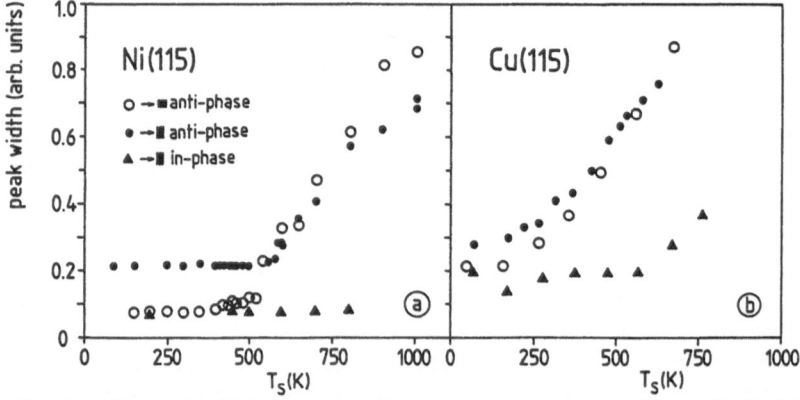

**Fig. 7.3.** He peak widths versus surface temperature measured for Ni(115) (a) [Conrad et al., 1986a] and for Cu(115) (b) [Fabre, Gorse, Salanon, Lapujoulade, 1987b]. Data were taken under in-phase conditions along the [$\overline{5}52$] azimuth (▲) and under anti-phase conditions along the [$\overline{5}52$] azimuth (•) and the [$1\overline{1}0$] azimuth (o), respectively

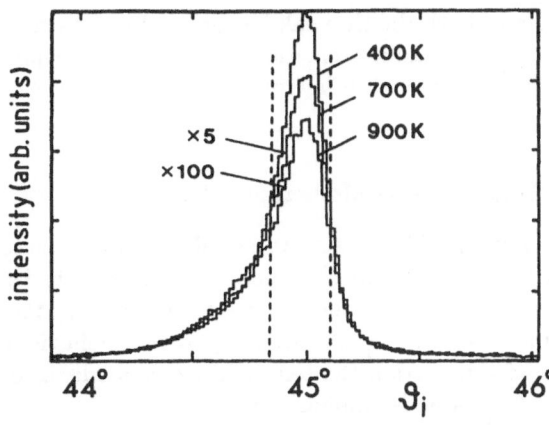

**Fig. 7.4.** He specular peak profile, obtained under out-of-phase conditions ($E_{He} = 18.3\,meV$, $\vartheta_i = 45°$) on Cu(011) at various surface temperatures $T_s = 400, 700$ and $900\,K$ [Zeppenfeld, Kern, David, Comsa, 1989]

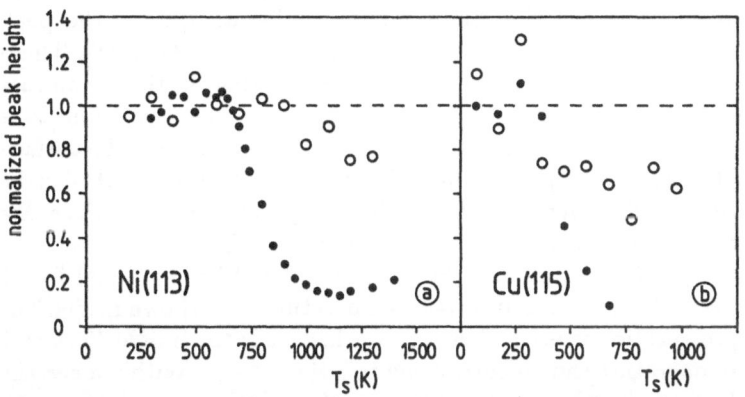

**Fig. 7.5.** He specular peak heights (corrected for Debye–Waller effects) vs surface temperature for Ni(113) (a) [Conrad, Allen, Blanchard, Engel, 1987] and Cu(115) (b) [Fabre, Gorse, Salanon, Lapujoulade, 1987b]. Data taken under in-phase (o) and anti-phase conditions (•), respectively

Lapujoulade, 1987b], respectively. Data are shown for both in- and anti-phase scattering conditions. All data have been corrected for Debye–Waller effects in the harmonic approximation by assuming that the squared vibrational amplitude increases proportional to the temperature. The behaviour of the in- and anti-phase peak heights with increasing surface temperature is quite different. The anti-phase peak heights drop dramatically at high temperatures, whereas the in-phase ones decay only slightly. As demonstrated in Chap. 6 the difference between the in- and anti-phase peak heights is due to the broadening of the anti-phase peaks, i.e. to the appearance of new reflecting terrace domains. The decrease of the in-phase peak heights is a measure of the diffuse scattering from the corresponding new domain boundaries. Quantitative evaluation is not yet feasible since both the cross-section for diffuse scattering at the domain boundaries and the exact correction for Debye–Waller effects are not well-known.

Again in contrast with the case of (11m)-nominally stepped surfaces, the in- and anti-phase peak heights of the Cu(011) surface with increasing temperature are indistinguishable [Zeppenfeld, 1988].

Summarizing, the measurement of both the He-peak heights and widths consistently evidence the occurrence of a roughening transition on nominally stepped Cu- and Ni-surfaces and no similar effect on the low-index Cu surfaces up to 900 K.

### 7.2.2 Peak Profile Analysis at Large Parallel Momentum Change

A full analysis of the anti-phase peak profiles is able to supply more complete information on the surface structure (see Chap. 6). Therefore, the elastic He intensity has been investigated at large He parallel momentum change, $\Delta k_\parallel$. First the dependence of the He intensity on $\Delta k_\parallel$ and its temperature behaviour are discussed. In the second part of this section the temperature dependence of the He intensity for a fixed large $\Delta k_\parallel$ value is emphasized.

Starting from more sophisticated theoretical considerations on statistical mechanisms, a roughening criterion based on the profile analysis of the peak wings has been developed [Den Nijs, Riedel, Conrad, Engel, 1985; 1986]. These authors demonstrate that the elastic He intensity in the wings of the He diffraction peaks, should decay with parallel momentum change $\Delta k_\parallel$ according to a power law. The absolute value of the exponent should decrease with the surface temperature $T_s$. The exponents pass through the value $-1$ when the height-height correlation logarithmically diverges with the distance parallel to the surface. By definition, at this point the roughening temperature is reached.

Experimental TEAS data, obtained by Conrad, Allen, Blanchard and Engel [1987] for Ni(113) in a broad range of surface temperatures, are shown in Fig. 7.6. A double-log representation is used; thus a power law leads to straight lines and the roughening temperature should correspond to a slope "$-1$". Within a certain window, limited at small $\Delta k_\parallel$ by the finite transfer width of the instrument and at large $\Delta k_\parallel$ by half of the zone boundary $2\pi/a_0$ the double-log plots are indeed linear over more than one order of magnitude. This applies for data taken both perpendicular (Fig. 7.6) and parallel to the steps. In the particular case of

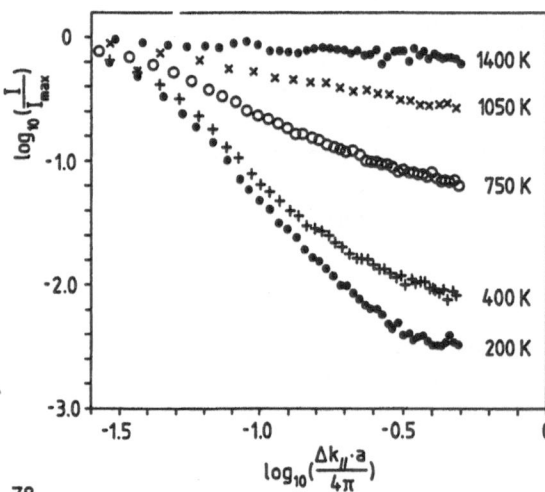

Fig. 7.6. Specular He peak profiles normalized to the peak maximum, measured at various temperatures of the Ni(113) surface. The data were taken along the [$\overline{3}32$] azimuth under anti-phase conditions ($E_{He} = 21\,meV$) [Conrad, Allen, Blanchard, Engel, 1987]

Ni(113) the roughening temperature $T_R$ amounts to $750 \pm 50\,K$ [Conrad, Allen, Blanchard, Engel, 1987]. The same procedure applied to Ni(115) results in $T_R = 450 \pm 50\,K$ [Conrad et al., 1985; 1986]. A similar ratio for $T_R$ has been found for the Cu(113) and -(115) surfaces; a detailed analysis of the He line shape leads to $T_R = 720 \pm 50\,K$ [Fabre, Salanon, Lapujoulade, 1988a] and $T_R \approx 380\,K$ [Fabre, Gorse, Salanon, Lapujoulade, 1987a; 1987b], respectively. These results are in accordance with recent Monte Carlo calculations [Salanon, Fabre, Lapujoulade, Selke, 1988].

The consistency of the results of Lapujoulade et al. for stepped copper and of Engel et al. for stepped nickel surfaces with respect to $T_R$ is remarkable. Controversy still exists, however, with respect to the azimuthal dependence of the anti-phase peak widths. Conrad, Allen, Blanchard, Engel [1987] emphasize that above $T_R$ the anti-phase peak width measured parallel to the steps is identical to that perpendicular to the steps. In contrast, Fabre, Gorse, Salanon, Lapujoulade [1987b] note that their peak profiles are anisotropic. This feature has recently been demonstrated most clearly in the case of Cu(1,1,11) [Fabre, Salanon, Lapujoulade, 1988b]. These authors interpret this observation as the consequence of the step-step repulsion. From the analysis of their data on Cu(115) and Cu(1,1,11) in terms of the VGL model, Fabre, Salanon and Lapujoulade concluded that the step-step repulsive energy scales inversely proportional to the fourth power of their mutual distance.

Again in contrast to the nominally stepped surfaces, the low indexed Cu(011) surface does not exhibit the characteristic feature of roughening (slope "$-1$") upon using peak analysis. Figure 7.7 shows the elastic He intensity as a function of the parallel momentum change, taken at various surface temperatures for Cu(011) by Zeppenfeld, Kern, David, Comsa [1989]. Note, that the plotted He intensities are truly *elastic* intensities, the data being taken in the high resolution time-of-flight mode; this is necessary for a rigorous comparison with the theoretical predictions. These log-log plots are as predicted [Den Nijs et al., 1986; Villain, Grempel, Lapujoulade, 1985] also linear; however, their slopes vary only slightly

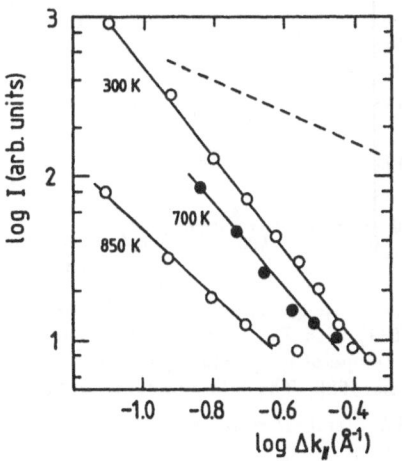

Fig. 7.7. Profiles of the purely elastic component of the He specular peak measured at various temperatures of the Cu(011) surface. The data are taken with $E_{He} = 18.3\,meV$ along the [001] azimuth [Zeppenfeld, Kern, David, Comsa, 1989]. The dashed line with the slope $-1$ would correspond to surface roughening [Conrad, Allen, Blanchard, Engel, 1987]

with temperature from $-2.8$ for $T_s = 300\,\mathrm{K}$ to $-2.0$ for $T_s = 850\,\mathrm{K}$ and do not approach the value $-1$, indicative of a roughening transition, up to $T_s = 850\,\mathrm{K}$.

The more complete He-peak profile analysis leads thus to the same conclusion: the nominally stepped surfaces [Cu(113), Cu(115), Cu(1,1,11), Ni(113) and Ni(115)] do roughen, while the low-index Cu(011) surface does not exhibit a behaviour characteristic for roughening, at least up to $T_s = 850\,\mathrm{K} - 900\,\mathrm{K}$.

Gorse and Lapujoulade [1985] suggested that pointlike defects and/or narrow ($< 10\,\text{Å}$ wide) terraces may proliferate at high temperature. Such behaviour can be traced by measuring the diffuse elastic He intensity at large parallel momentum change. It has been shown in preceding chapters that individual diffuse scatterers [e.g. CO/Pt(111), monovacancies, or step rows induced by ion bombardment] lead to a general increase of the He diffuse elastic intensity. Of course, the onset of a roughening transition will also lead to an increased diffuse elastic scattering. This is illustrated in Fig. 7.8 showing the diffuse elastic scattering for Ni(113) as a function of surface temperature [Conrad, Allen, Blanchard, Engel, 1988]. The diffuse elastic He intensity, corrected for Debye–Waller effects, *increases* above about 600 K in accordance with the increase of the length of the diffusely scattering domain boundaries due to thermal roughening. For comparison, Fig. 7.9 shows the diffuse elastic He intensity vs surface temperature taken at $\Delta k_\parallel = 0.73\,\text{Å}^{-1}$ for Cu(011) [Zeppenfeld, Kern, David, Comsa, 1989]. The data (also corrected for Debye–Waller effects) shows no sign of an increase of the diffuse elastic He intensity up to the highest temperatures. The conclusion is thus the same as in the preceding section: the nominally stepped surfaces roughen, while the low-index ones apparently do not.

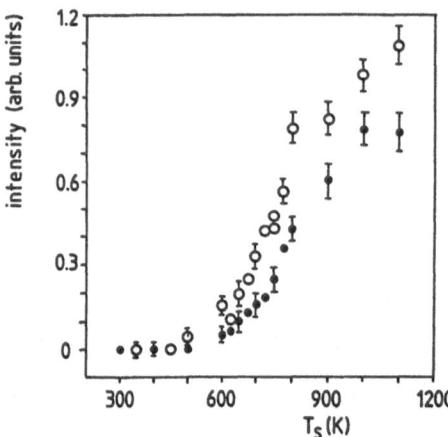

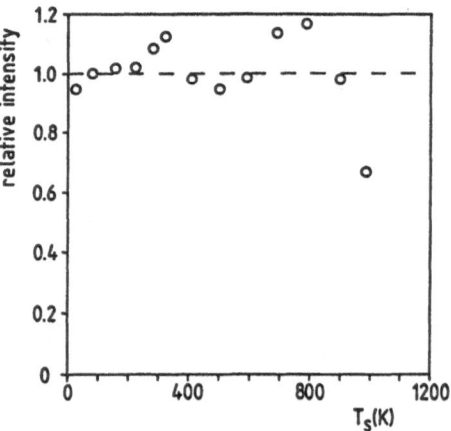

Fig. 7.8. Diffuse elastic scattering of He from Ni(113) with $E_{\mathrm{He}} = 21\,\mathrm{meV}$ versus surface temperature [Conrad, Allen, Blanchard, Engel, 1988]. The data taken at $k_\parallel = 1.5\,\text{Å}^{-1}$ ($\bullet$) and $1.0\,\text{Å}^{-1}$ (o) are corrected for Debye–Waller effects

Fig. 7.9. Relative diffuse elastic He intensity Cu(011) with $E_{\mathrm{He}} = 18.3\,\mathrm{meV}$ versus surface temperature [Zeppenfeld, Kern, David, Comsa, 1989]. The data were taken at $\Delta k_\parallel = 0.73\,\text{Å}^{-1}$ and are corrected for Debye–Waller effects

## 7.3    The Anomalous Behaviour of the Specular He Peak Height from Cu(011) Versus Surface Temperature

As inferred from the discussions in the previous sections, the low-index faces of copper do not appear to undergo a roughening transition. In this section we address the question how to reconcile this fact with the observation of the anomalous decrease of the He specular intensity with surface temperature (see Fig. 7.1). Following a suggestion of Armand, Gorse, Lapujoulade, Manson [1987], the experimental features on these surfaces may be attributed to anomalous inelastic effects due to anharmonicities of the surface atom potential (see also Jayanthi, Armand, Manson [1985]). If so, the mean square vibrational amplitudes of the surface atoms should no longer be proportional to $T_s$ as is the case in the harmonic approximation.

In order to investigate this assumption the He specular peak heights have been expressed in terms of effective mean square vibrational amplitudes of the surface atoms $\langle u_{\perp}^2 \rangle$, using the well-known Debye–Waller expression [Levi, Suhl, 1979]. The occurrence of surface anharmonicity effects would lead to $\langle u_{\perp}^2 \rangle$ increasing more strongly than proportional to $T_s$. Figure 7.10 shows the effective mean square vibrational amplitudes for Cu(011) and Cu(001) versus the surface temperature in the frame of the above assumption.

Note that the effective surface vibrational amplitude, as inferred from the height of thermal He peaks versus temperature, is influenced by the "Armand-effect", i.e. by the simultaneous interaction of the probing He atom with several surface atoms [Armand, Lapujoulade, Lejay, 1977]. This feature complicates the evaluation; for instance lesser correlation between the vibrations of neighbouring atoms leads to larger effective mean square amplitudes.

The data shown in Fig. 7.10 for Cu(011) and Cu(001) show clear similarities. First, i.e. at lower temperatures, the effective mean square amplitude increases linearly with temperature in accordance with the harmonic approximation. At higher temperatures the effective mean square vibrational amplitudes increases

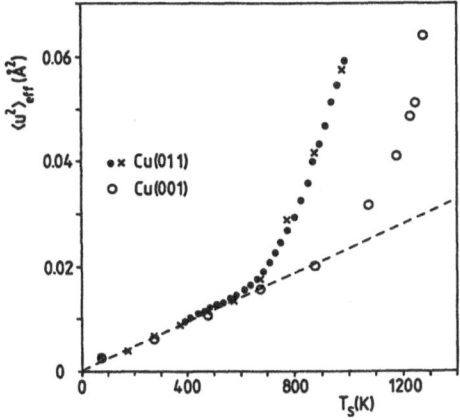

Fig. 7.10. Effective mean square vibrational amplitudes of the surface atoms (see text) of Cu(001) (o; [Armand, Gorse, Lapujoulade, Manson, 1987]) and of Cu(011) (×, [Gorse, 1985]; •, [Zeppenfeld, Kern, David, Comsa, 1989]) as a function of the surface temperature

faster with temperature, probably due to surface anharmonicity effects. The anomalies occur at about 300 K lower temperatures in the (011) than in the (001) case. This may be attributed either to a different temperature dependence of the "Armand-effect", e.g. due to the lack of nearest neighbours perpendicular to the close-packed rows, or to stronger anharmonicity on the (011) face than on the (001) face. The latter is consistent with recent observations on the Pb(011) surface. In contrast to other low-index Pb surfaces the (011) face shows a substantial relaxation; moreover the relaxation on Pb(011) depends strongly on temperature, a fact that is consistent with surface anharmonicity [Frenken, Huussen, Van der Veen, 1987].

In conclusion, the investigation of the He specular peak widths and heights, as well as of He diffuse elastic scattering, allows an accurate and detailed classification of surface temperature effects. The simple evaluation of the coherent scattered intensity of thermal He atoms, of low energy electrons or of X-rays is insufficient.

# 8. TEAS as a Probe of Surface Coverage

TEAS exhibits a unique combination of capabilities for monitoring the surface coverage: extreme sensitivity and complete inertness. The first is due to the very large cross-sections of the adparticles for He diffuse scattering (Chap. 3), the latter to the fact that the noble gas helium atom at these low energies, $10-100\,\mathrm{meV}$, is both chemically and physically completely inert. These properties overcompensate in many cases the inability to determine the nature of the adsorbed species. Because of this inability it is always necessary to work under well-defined conditions in order to know the nature of the adsorbing species. This ought to be anyhow a matter of course in modern surface science. The case examples presented in the subsequent sections show a great richness of information obtained with an accuracy hardly accessible with other techniques.

As usual, for getting more and higher quality information, one has to pay more: one has to know quantitatively the dependence of the He relative reflectivity, $r = I/I_0$, on the coverage, $\Theta$, for the considered system under the given experimental conditions. As discussed in the preceding chapters this dependence is determined by the nature and size of the scattering cross-section of the individual adsorbates and by their actual distribution.

The actual shape of the dependence $r(\Theta) = I(\Theta)/I_0$ may be of course obtained for each particular system fully empirically. The He reflectivity is monitored continuously as a function of exposure; then, the exposure is transformed into a coverage scale by measuring the coverage corresponding to a number of well-defined exposures by thermal programmed desorption (TPD), or any other faithful method. The calibration curve $r(\Theta) = I(\Theta)/I_0$ can then be used for detailed continuous and non-damaging TEAS measurements, down to very low coverages ($\approx 0.001$).

It is certainly much more efficient and elegant to use an analytical function for the $r(\Theta) = I(\Theta)/I_0$ dependence. In previous chapters we have deduced on the basis of the "overlap approach" a number of analytical forms $r(\Theta) = I(\Theta)/I_0$ applicable for various types of scatterers with various lateral distributions, e.g.: lattice gas of purely diffuse scatterers with large cross-sections (3.7) and with smaller ones (3.9); lattice gas of purely diffuse scatterers with large cross-sections and exclusion of nearest neighbours (4.5); strongly repulsive, purely diffuse scatterers (4.2); islands of strongly attractive purely diffuse scatterers (4.3 and 4.4); lattice gas of partially reflecting scatterers with relatively small cross-sections for diffuse scattering (3.9 – 11). Upon a detailed analysis of the scattering behaviour and of the lateral distribution, most systems considered so far could be associated with one or the other of these expressions. The parameters entering these expressions, like the cross-section for diffuse scattering $\Sigma$ and, in the case of partially reflecting adsorbates, also the reflectivity, $\varrho$, and the "height" of the adlayer, $h$, were determined from the experiment. In all cases these $r(\Theta) = I(\Theta)/I_0$ expressions fit the data with remarkable accuracy. Examples have been given so far in Fig. 4.2 for CO/Pt(111) fitted by (4.5) and in Fig. 4.6 for Xe/Pt(111) fitted by (4.5) and (4.3). In both cases the adsorbates are purely diffuse scatterers so that only one parameter, the cross-section $\Sigma$, has to be fitted from the data. The case of hydrogen on Pt(111), a highly reflecting adsorbate, discussed in Sect. 8.1 below, is more complicated: four parameters have to be fitted from the data. However, because they are obtained from independent experiments, they are by no means arbitrary. The $r(\Theta) = I(\Theta)/I_0$ analytical functions obtained in this way are also calibration curves. In contrast to the purely empirical ones they have the practical advantage that — being based on the detailed knowledge of the scattering behaviour and of the lateral distribution of the adsorbates — they can be used under widely differing experimental conditions.

We will describe first in Sect. 8.1.1 the procedure followed to establish the analytical calibration function, $r(\Theta) = I(\Theta)/I_0$, in the most complicated case solved so far [H/Pt(111)] and its experimental check. In Sect. 8.1.2 we will exemplify the use of the calibration function to monitor the H coverage in a very demanding study of the adsorption of hydrogen on Pt(111). The use of the simple calibration functions associated with purely diffuse scatterers is illustrated in Sect. 8.2 by monitoring adsorption isotherms of CO on, and desorption transients of CO from Pt(111), as well as quasi-equilibrium isobars of Xe on the same surface. Finally, in Sect. 8.3 a highly accurate procedure to determine changes of the sticking probability with some parameter (e.g. the surface temperature) will be presented. In contrast to the other uses of TEAS as a probe for surface coverage, this latter application is independent of any calibration function.

## 8.1 The H/Pt(111) System

### 8.1.1 The Calibration Function for Highly Reflecting Scatterers

The characteristic shape of the hydrogen adsorption curve is illustrated in Fig. 8.1, by the double-logarithmic plot of the relative He reflectivity (relative specular peak height $I/I_0$) vs $H_2$ exposure of the Pt(111) surface at $T_s = 155\,\mathrm{K}$ ($E_{He} = 16\,\mathrm{meV}$, $\vartheta_i = \vartheta_f = 40°$). Similar to the case of "purely" diffuse scatterers (CO in Fig. 4.2 and Xe in Fig. 4.6) the He reflectivity decreases rapidly in the initial stages of adsorption. However, in contrast to that case the He reflectivity after passing through a minimum recovers at H saturation its initial clean surface value almost completely. Helium diffraction measurements show that the saturated hydrogen adlayer is largely uncorrugated: the first order diffraction peaks are two orders of magnitude less intense than the specular peak [Lee, Cowin, Wharton, 1983]. Thus the H-saturated surface can be viewed as a mirror for He scattering of almost the same quality like the clean Pt(111) surface; only their Debye-temperatures differ slightly: 230 and 190 K for the clean and the H satu-

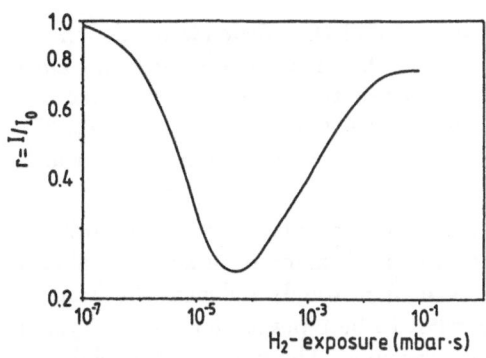

Fig.8.1. Relative He specular peak height during hydrogen adsorption on Pt(111) at $T_s = 155\,\mathrm{K}$ vs $H_2$ exposure ($E_{He} = 16\,\mathrm{meV}$; $\vartheta_i = 40°$)

rated Pt(111) surface, respectively [Lee, Cowin, Wharton, 1983; Poelsema et al., 1986].

The behaviour of the He reflectivity as a function of hydrogen coverage can be understood within the framework of the model ansatz discussed in Chap. 3. In the low H-coverage regime, the main contribution to the He specular beam comes from the scattering on the clean Pt areas, $A_S$, non-disturbed by the diffuse scattering at the H adatoms (cross-section, $F_A$); at high H coverages it comes from the scattering from H-saturated areas, $A_A$, non-disturbed by the diffuse scattering at H-adlayer vacancies (cross-section, $F_V$). At intermediate coverages the interference between the scattering amplitudes $A_S$ and $A_A$ in the specular direction plays an important role. The reflectivity can be described by

$$r = I/I_0 = (A/A_0)^2 = [A_S^2 + 2A_S A_A \cos\varphi + A_A^2]/A_0^2 \quad , \tag{8.1}$$

which is directly obtained from (3.11) and where $A_S/A_0$ and $A_A/A_0$ denote the relative scattering amplitudes given by equations of the type (3.9 and 10), respectively. The phase difference between scattering contributions into the specular direction from clean parts of the Pt surface and those of hydrogen covered parts, separated by a height difference $h$, is given by

$$\varphi = 2\pi\,\frac{2h\cos\vartheta_i}{\lambda} \quad , \tag{8.2}$$

where $\lambda$ and $\vartheta_i$ denote the wavelength and the angle of incidence of the probing He beam.

The relative He reflectivity vs coverage described by (8.1) is obviously the calibration curve for H/Pt(111) we are seeking. It contains [see (8.2, 3.9 and 10)] the four parameters $F_A = \Sigma_A/2$, $F_V = \Sigma_V/2$, $\varrho$ and $h$ which have to be determined from the experiment:

1) The parameter $\varrho$ is the reflectivity of the saturated H adlayer defined in Sect. 5.2.2:

$$\varrho = \left(\frac{A_A}{A_0}\right)^2\Bigg|_{\Theta_H=\text{sat}} = \frac{I}{I_0}\Bigg|_{\Theta_H=\text{sat}} = r(\Theta_H = \text{sat}) \quad .$$

Because the H-saturated surface is almost as uncorrugated as the clean Pt(111) surface, $\varrho$ is practically equal to the ratio of the Debye–Waller factors of the

H-saturated and of the clean Pt(111) surface and thus depends on the energy, $E_{He}$, on the surface temperature $T_s$ and on the incident and outgoing angle, $\vartheta_i = \vartheta_f$, of the He beam. Its value is obtained straightforwardly either from the adsorption curve or from the knowledge of the Debye-temperatures ($\theta_D^{clean} = 230\,K$, $\theta_D^{H=sat} = 190\,K$) and of the scattering conditions (change of the He perpendicular momentum).

2) The height difference $h$ between the two specularly reflecting "mirrors" — the H adlayer and the clean Pt surface — is obtained from a more elaborate experiment. However, because the value of $h$ is widely independent from the scattering conditions, the procedure has not to be repeated for a given system. The procedure is based on the following reasoning. The value of the height $h$ is obtained from (8.2) by determining experimentally the incidence angle $\vartheta_i(-\vartheta_f)$ for which a given phase difference $\varphi$ between the He waves scattered from the He adlayer and the Pt surface is obtained. It is obvious from (8.1) that the value of the reflectivity in the minimum of the adsorption curves $r_{min} = I/I_0|_{min}$ is mainly determined by the interference of the He waves scattered from the two "mirrors". For instance, for "in-phase" conditions ($\varphi = 2\pi n$) the largest value of $r_{min}$ is expected, while for antiphase conditions ($\varphi = 2\pi(n + 1/2)$) the smallest — a kind of "absolute minimum" — should be obtained. This is illustrated in Fig. 8.2: the values of the minima $r_{min} = I/I_0|_{min}$ of adsorption curves (like that in Fig. 8.1) are plotted as a function of the incident angle $\vartheta_i(= \vartheta_f)$ for He beams of two wave-lengths $\lambda = 0.57$ and $0.36$ Å ($E_{He} = 63$ and $160$ meV, respectively). The "absolute minimum" which is observed in each of the two curves at $\vartheta_i = 37.5°$ and $\sim 60°$, respectively, corresponds to the lowest order ($n = 0$) anti-phase condition. From (8.2) one gets $h = 0.180$ and $0.177$ Å, respectively. Because the height $h$ is much smaller than the wavelength $\lambda$, not even the lowest nontrivial in-phase condition, $n = 1$, can be observed: the corresponding maximum in the $r_{min}$ vs $\vartheta_i$ curve should be located at $\vartheta_i = 0°$, for $h = 0.18$ Å and $\lambda = 0.36$ Å (see 8.2), i.e. out of the experimental range. Note that the height $h = 0.18$ Å is of the same order as the H-induced corrugation of $0.23$ Å on Ni(110) as obtained from He diffraction experiments [Engel, Rieder, 1982].

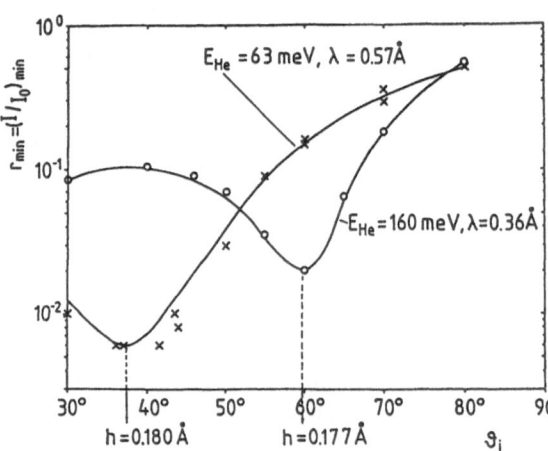

Fig. 8.2. The value of the minimum reflectivity, $r_{min}$, of H/Pt(111) adsorption curves like that shown in Fig. 8.1 vs the angle of incidence of the He beam. The data were taken at $E_{He} = 63$ meV ($\times$) and $160$ meV (○)

3) The estimation of the values of the two cross-sections for diffuse scattering $F_A = \Sigma_A/2$ and $F_V = \Sigma_V/2$ requires, as in the case of purely diffuse scattering, the transformation of the exposure, $\varepsilon$, into the coverage, $\Theta$, scale. In the simpler case of isolated purely diffuse scatterers (see, e.g., Sect. 3.3) it is sufficient to perform the transformation at low coverages in order to get the *initial* slope of the $r(\Theta) = I(\Theta)/I_0$ curve; it is thus in principle sufficient to know the initial sticking coefficient. In the case of highly reflecting adsorbates the transformation has to be done at low coverages and at coverages near saturation, where the He scattering is dominated by $F_A$ and $F_V$, respectively (compare Sect. 3.5.2). In view of the complexity introduced by the He wave interference it is more accurate to determine experimentally the relation between coverage and exposure in the entire coverage range $0 \leq \Theta_H \leq 1$. The result is a calibration plot as shown in Fig. 8.3. The data points are obtained by thermal programmed desorption (TPD) upon numerous hydrogen exposures at $T_s = 80$ K.

In the next step we use the calibration data points in Fig. 8.3 to transform adsorption curves $r(\varepsilon) = I(\varepsilon)/I_0$ like that in Fig. 8.1 into $r(\Theta) = I(\Theta)/I_0$ curves. This has been done for two adsorption curves measured at $T_s = 80$ K with an $E_{He} = 63$ meV He beam incident at $\vartheta_i = 37.5°$ and $66°$ (Fig. 8.4). We have arbitrarily chosen these two scattering conditions, corresponding to $\varphi = \pi$ and $\varphi = \pi/2$, because they lead to particularly simple shapes of (8.1):

$$ r = I/I_0 = (A_S - A_A)^2/A_0^2 \quad \text{and} \quad r = I/I_0 = (A_S^2 + A_A^2)/A_0^2 \quad , $$

respectively. Now, from the initial and final slope of the data in Fig. 8.4, and assuming that both the H adatoms at $\Theta_H \approx 0$ and the vacancies in the nearly saturated H adlayer (at $\Theta_H \approx 1$) have a lattice gas distribution, a rough estimate of $F_A$ and $F_V$ is obtained: around $10\,\text{Å}^2$. In view of this size ($\eta \approx \xi = \sqrt{Fn_s} \approx 1.23 < 3/2$; see Sects. 3.5.1 and 2) the expressions (3.9 and 10) can be used to calculate $A_S/A_0$ and $A_A/A_0$ needed in the calibration function (8.1). The optimal

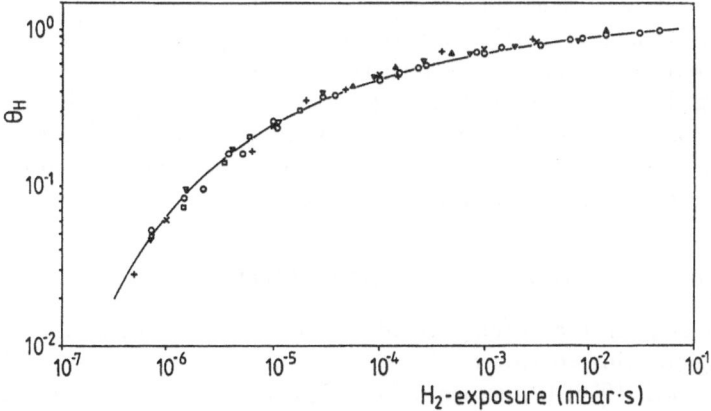

**Fig. 8.3.** Calibration of the H adsorption on Pt(111) at $T_s = 80$ K. Data are taken from several TPD series ($\times, +, \circ, \Delta, \nabla,$  ) and from one TEAS experiment (*solid curve*); see text

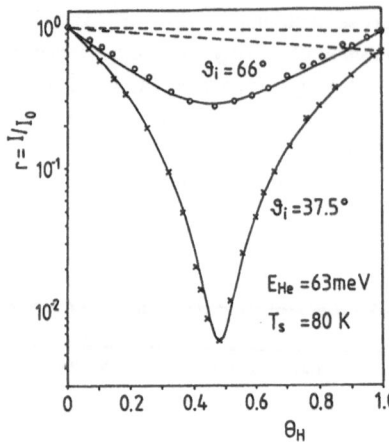

**Fig. 8.4.** Relative He specular peak height vs hydrogen coverage, measured at $\vartheta_i = 37.5°$ (×) and $\vartheta_i = 66°$ (o) respectively. The solid curves are calculated using the fitted parameters and assuming a random distribution of hydrogen adatoms and vacancies (see text). For comparison the dashed curves show the expected behaviour in the case of large hydrogen and vacancy islands

values of the cross-sections $F_A$ and $F_V$ are obtained now by means of an iterative procedure. The cross-section values leading together with the parameters $\varrho$ and $h$ determined as described above, to the best fit of the $\varphi = \pi$ data (lower curve in Fig. 8.4) are $F_A = 12\,\text{Å}^2$ and $F_V = 9\,\text{Å}^2$. The same fit quality is obtained also for very different interference conditions as illustrated for the upper data points in Fig. 8.4 obtained for $\varphi = \pi/2$. This demonstrates that (8.1) (together with (8.2), the lattice gas equations (3.9 and 10) and the set of parameters $\varrho$, $h$, $F_A$ and $F_V$) represents an excellent calibration function for the system H/Pt(111). (The importance of the lattice gas assumption is illustrated by the dashed lines in Fig. 8.4; these lines represent a "fit" to the data by assuming the formation of large islands.)

The relatively laborious determination of the four parameters has to be done only once for a given system. The calibration function can thus be used to monitor continuously, in detail and non-damagingly the H coverage, a task not solved by any other method so far. An application of this capability of TEAS is shown in the next section. In order to get a feeling for the accuracy of the calibration function derived here, the experimental correlation between H coverage and $H_2$ exposure determined now from monitoring the He relative reflectivity ($r = I/I_0$) vs $H_2$ exposure and using the calibration function is plotted as a solid line in Fig. 8.3. Note that from the slope of this adsorption curve the $H_2$ sticking probability vs H coverage, $s_{H_2}(\Theta_H)$, can be directly obtained. This procedure has been used throughout the investigations described in the next section.

### 8.1.2 The Dissociative Adsorption of Hydrogen on Pt(111)

The role of atomic steps in the dissociative adsorption of hydrogen on Pt(111) is a matter of long-standing controversy in the literature [Bernasek, Somorjai, 1975; Christmann, Ertl, 1976; Christmann, Ertl, Pignet, 1976; Wachs, Madix, 1976; Collins, Spicer, 1977a; 1977b; Salméron, Gale, Somorjai, 1977; 1979]. After an initial claim that the ratio of the dissociative sticking probabilities at step sites to that at (111)-terrace sites is $10^3 - 10^4$ [Bernasek, Somorjai, 1975] this value stabilized around 10. The present TEAS investigation leads to the result that, at least at low surface temperatures, the initial claim was correct, i.e. the dissociation of thermal energy $H_2$ molecules takes place exclusively at step sites and not at the (111) terraces. The fact that this surprising conclusion has been obtained now on a system investigated thoroughly in the past is due primarily to two outstanding capabilities of TEAS: (a) the continuous and precise monitoring of the H coverage discussed in the preceding section and (b) the new procedure — developed by means of TEAS and described in Sect. 6.5.3 — to create and characterize in situ surfaces with step densities in the range $0.001 - 0.15$. The demonstration leading to the conclusion above has to be circumstantial because even at step densities of 0.001 the $H_2$ dissociation is still substantial and because lower step densities have not been achieved so far on Pt surfaces. The demonstration is also rather cumbersome, so that we will give here only some of the main experimental findings which illustrate the capabilities of TEAS.

In Fig. 8.5, we show the dependence of the dissociative sticking probability of hydrogen, $s_{H_2}$, at $T_s = 155\,K$ on the H coverage, $\Theta_H$, with the step density as parameter. In order to evidence a possible Langmuir-adsorption behaviour, not $s_{H_2}$ but $s_{H_2}/(1 - \Theta_H)^2$ is plotted on the ordinate: a Langmuir behaviour would show up as $s_{H_2}/(1 - \Theta_H)^2 = $ constant. From the inspection of the data in Fig. 8.5 it appears that each curve is made up of two distinct regions: a quasi-exponential decay and a "Langmuir" region in the higher $\Theta_H$ range. The weight of the latter increases with the step density. The exponential behaviour can hardly be explained within the frame of the models proposed so far, in particular within a barrier model.

The model developed on the basis of the TEAS investigation illustrated here assumes that the dissociative adsorption takes place exclusively at step sites which can be reached along two distinct channels: (1) $H_2$ molecules incident on ideal terrace sites are adsorbed with a certain probability in a *molecular* state;

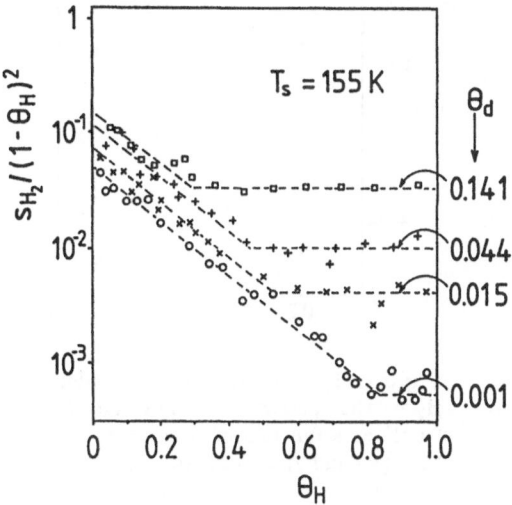

**Fig. 8.5.** Dissociative sticking probability $s_{H_2}$ of $H_2$ on Pt(111), plotted as $s_{H_2}(1 - \Theta_H)^{-2}$, vs hydrogen coverage, $\Theta_H$. All data are taken at $T_s = 155\,K$ on the same Pt crystal upon in situ preparation of step concentrations in the range $\Theta_d = 0.001 - 0.141$ by high temperature ion bombardment

these admolecules migrate across the surface until they either desorb or encounter a step site; (2) $H_2$ molecules incident near a step site. In either case, the $H_2$ molecules present near a step site may dissociate with a certain probability, if there are two neighbouring empty sites available for the two H atoms (Langmuir adsorption).

The channel (2) corresponds obviously to the Langmuir region in Fig. 8.5. According to the model, the dissociative adsorption probability through channel (2) has to be proportional to the probability of finding two neighbouring empty sites, i.e. to $(1 - \Theta_H)^2$, and to the density of step sites. Both conditions are confirmed by the data in Fig. 8.5 in the "Langmuir region": The ratio $s_{H_2}/(1 - \Theta_H)^2$ is constant and the sticking probability is proportional to the step density, $\Theta_d$. The latter can be seen in Fig. 8.6, where the sticking probability in the Langmuir region, $s_L$, obtained from Fig. 8.5 is plotted vs $\Theta_d$. The $s_L$ points lie on a straight line which — within the experimental accuracy — passes through the origin. The Langmuir adsorption process is obviously negligibly small on the ideal (111) terraces.

The dissociative adsorption via channel (1) leads to the quasi-exponential behaviour at lower $\Theta_H$ observed in Fig. 8.5. There are two possible mechanisms which qualitatively explain this behaviour. Both are related to the probability of the $H_2$ molecules trapped in a molecular state to encounter a step site *before* desorbing. The binding energy in the molecular state might decrease (linearly) with the H coverage $\Theta_H$; this would lead to an exponential decrease of the life-time in the molecular state with $\Theta_H$ and thus also of the probability to encounter a step site. Another possibility is that, due to collisions with H adatoms, the time needed by $H_2$ in the molecular state to encounter a step site increases correspondingly. It has yet not been possible to distinguish between these mechanisms, but the basic idea — i.e., the dependence on the life-time in the molecular state — could be verified by increasing drastically the molecular state life-time by cooling the surface. This is illustrated in Fig. 8.7. The $s_{H_2}$ vs $\Theta_H$ data are obtained on the

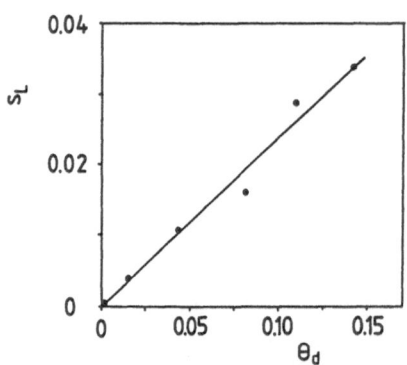

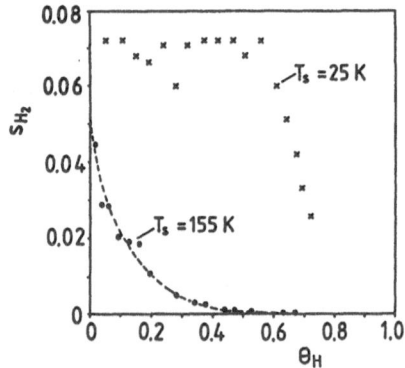

**Fig. 8.6.** Dissociative sticking probability for $H_2$ on Pt(111) through the "Langmuir" channel, $s_L$ (see text and Fig. 8.5) vs step density

**Fig. 8.7.** Dissociative sticking probability of $H_2$ on the "perfect" Pt(111) surface ($\Theta_d \approx 0.001$) as a function of H coverage at $T_s = 25\,K$ (×) and $T_s = 155\,K$ (•)

Pt(111) surface with the lowest density of steps ($\Theta_d \approx 0.001$). The data points measured at $T_s = 155\,\text{K}$ are taken from Fig. 8.5, but plotted here on a linear $s_{H_2}$ scale. The shape of the $s_{H_2}$ vs $\Theta_H$ dependence at $T_s = 25\,\text{K}$ is dramatically different: the sticking probability stays constant up to $\Theta_H \approx 0.6$ and is $1.5 - 500$ (depending on $\Theta_H$) times larger than at $155\,\text{K}$. This behaviour is obviously at odds with any type of barrier model. However, it can be straightforwardly understood on the basis of the present model: at $T_s = 25\,\text{K}$ the life-time of $H_2$ in the molecular state is up to $\Theta_H \approx 0.6$ sufficiently long for *all* H molecules trapped in this state to encounter a step site before desorbing. This is a strong argument in favour of this model. Additional arguments obtained from other types of TEAS and TPD experiments further strengthen the model assumption [Lenz, 1987], but would expand the discussion here, aimed to illustrate the TEAS capabilities, too much.

## 8.2 Equilibrium and Desorption Experiments with Purely Diffuse Scatterers

In the case of purely diffuse scatterers the intensity of the specular beam originates, by definition, exclusively from the specular reflection on the substrate areas not covered by the cross-sections of the scatterers. As discussed in this and previous chapters, this leads in general to a very simple calibration function $r(\Theta) = I(\Theta)/I_0$. It consists only of the first term of (8.1) and contains only one parameter, the cross-section for diffuse scattering, $\Sigma$. The derivation of an adequate calibration function is obviously much easier than in the case of the H/Pt(111) system. The practical use of the calibration function when monitoring the coverage is also much simpler even if the system parameters, in particular the surface temperature, are varied in a large range, unless the lateral distribution would change. Indeed, except for extremely low temperatures ($\leq 20\,\text{K}$) where their specular scattering contribution is no more negligible (even within the modest dynamical range limits set in the present discussion – see Sects. 2.2 and 3.1) and thus they are no more purely diffuse scatterers, the admolecules investigated so far (CO, NO, Xe, Kr, $H_2O$) do not change their cross-sections with the surface temperature, see e.g. [Poelsema, Verheij, Comsa, 1984b]; there is in fact no apparent reason for such a change. As long as the admolecules are purely diffuse scatterers the specular intensity, $I$, measured in the presence of a certain coverage $\Theta$ originates exclusively in the He specular scattering from the substrate areas not covered by the adsorbate. These areas have the same reflectivity per unit area (i.e. the same Debye–Waller factor) as the perfectly clean substrate, $\Theta = 0$. (It has been, indeed, proven in a few cases that the presence of adsorbates does not change the vibrational properties, i.e. the Debye-temperature, of the neighbouring, not covered areas [Poelsema, Mechtersheimer, Comsa, 1981b].) Therefore, the intensities $I$ and $I_0$ are affected by *the same* Debye–Waller factor and thus in the ratio $I(\Theta)/I_0$ the Debye–Waller factor is automatically cancelled out. The calibration function $r(\Theta) = I(\Theta)/I_0$ becomes an unequivocal function of coverage, independent of temperature; it is simply the fractional area not covered by the scattering cross-sections of the admolecules. This conclusion appeared to be very useful in the experiments shown below.

### 8.2.1 Equilibrium Experiments

Experiments under equilibrium conditions for the determination of thermodynamic quantities play an important role in surface science. (Rigorously, the experiments are conducted in "quasi"-equilibrium since in general the surface and gas temperatures differ.) Three quantities are controlled in these experiments: the 3D-gas pressure, the surface temperature and the adsorbate coverage. We will give here examples in which the coverage is monitored by TEAS. Depending on which parameter is kept constant, one obtains isotherms ($T_s$ = constant), isobars ($p_{3D}$ = constant) or isosteres ($\Theta$ = constant). It is obviously sufficient to measure any family of these curves; the other two families can then be inferred straightforwardly. Thus the particular choice depends on the given conditions of the experiment. It is for instance often easy to maintain $T_s$ = constant and to monitor the He-specular peak height $I$ vs $p_{3D}$. The value of $I_0$ at the given $T_s$ is obtained for $p_{3D} = 0$. By means of the calibration function, $I/I_0$ is transformed into the coverage $\Theta$. The isotherm, i.e. $\Theta$ vs $p_{3D}$ at $T_s$ = constant, can then be plotted directly. The measurement of isobars or isosteres is in principle not more complicated. One has only to take care to divide each value of $I$ by $I_0$ measured at the *same* $T_s$ in order to cancel the Debye–Waller factor properly. Two remarks are here in order:

i)   The isobars are preferred when the equilibrium pressure exceeds $\approx 10^{-6}$ mbar, i.e. when the attenuation of the He beam by gas phase scattering is no longer negligible. Since each isobar, $I$ vs $T_s$, is measured at $p_{3D}$ = constant, i.e. at constant attenuation, its shape is not influenced by the attenuation.

ii)  The dependence $I/I_0$ vs $\Theta$ being independent of $T_s$ for purely diffuse scatterers, the isosteres are in fact $I/I_0$ = constant curves. Thus the isosteric heat of adsorption obtained via the Clausius–Clapeyron equation from a given isostere does not imply the knowledge of the calibration function $I(\Theta)/I_0$; its value is thus independent of any model. The calibration function is necessary only to determine the coverage at which the heat of adsorption has been estimated. Experiments involving isotherms and isobars are illustrated by subsequent examples.

### 8.2.2 Isotherms

Measured isotherms $I/I_0(p_{CO})\big|_{T_s=\text{constant}}$ for CO/Pt(111) are shown in Fig. 8.8. The adsorption of CO is non-dissociative; the equilibrium expression is of the form [Poelsema, Palmer, Comsa, 1982; Verheij et al., 1987; Ibach, Erley, Wagner, 1980]

$$p_{CO} = \frac{\Theta}{1-\Theta} \cdot K(\Theta) \cdot T_s^3 \cdot \exp\left[-\frac{E_d(\Theta)}{kT_s}\right] \quad , \tag{8.3}$$

where the factor $K(\Theta)$ contains both the sticking probability and the frequency factor for desorption and $E_d(\Theta)$ represents the activation energy for desorption; both $K$ and $E_d$ may depend on coverage. The ordinate of the isotherms in Fig. 8.8 can be transformed in coverage by means of the calibration function (4.5) with $\Sigma = 123\,\text{Å}^2$. From each isotherm a semi-log plot of $[p_{CO}(1 - \Theta_{CO})/\Theta_{CO}]_{T_s}$ as a function of $\Theta_{CO}$ is obtained and shown in Fig. 8.9. These curves yield straightforward information on the coverage dependence of $K$ and $E_d$. In view of (8.3), the straight lines in Fig. 8.9 suggest two simplifying alternatives: either (a) K depends exponentially on $\Theta_{CO}$ and $E_d$ is a constant or (b) K is cover-

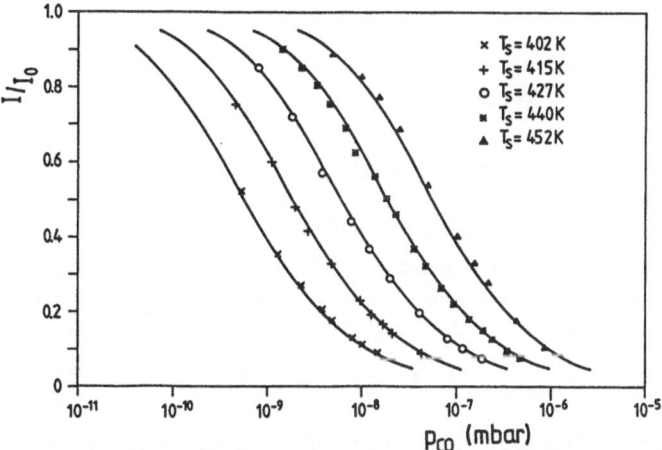

**Fig. 8.8.** Isotherms, i.e. the relative He specular peak height as a function of the CO equilibrium pressure for CO/Pt(111). ($E_{He} = 63\,\text{meV}$; $\vartheta_i = 40°$). The data are obtained at various surface temperatures $T_s$: 402 K ($\times$), 415 K (+), 427 K (o), 440 K ($\star$) and 452 K ($\blacktriangle$). The solid curves correspond to (8.3) (see text)

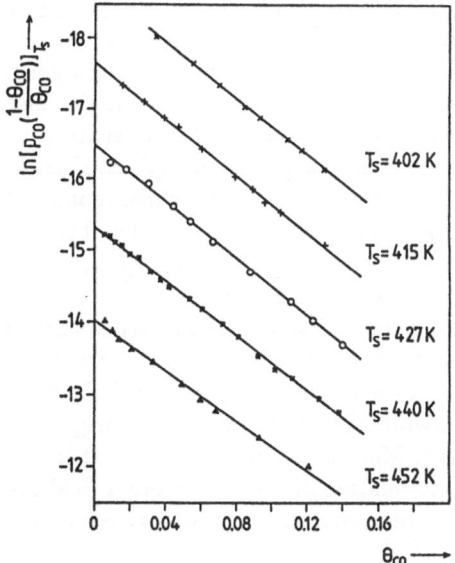

**Fig. 8.9.** Replot of the isotherms in Fig. 8.8 as $\ln[p_{CO}(1-\Theta_{CO})/\Theta_{CO}]_{T_s}$ versus CO coverage, $\Theta_{CO}$

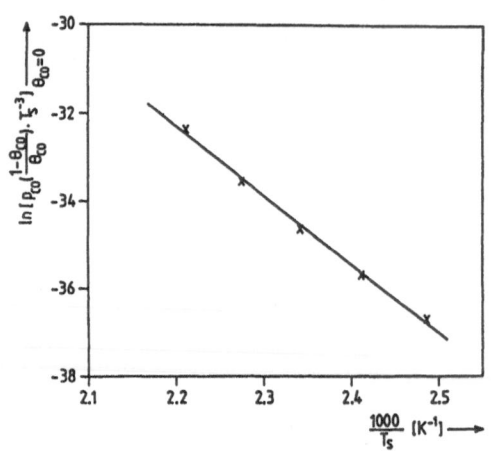

**Fig. 8.10.** Semilog plot of
$$\ln\left\{[p_{CO}(1 - \Theta_{CO})/\Theta_{CO}]T_s^{-3}\right\}_{\Theta_{CO}=0},$$
as deduced from Fig. 8.9, versus the reciprocal surface temperature

age independent and $E_d$ depends linearly on $\Theta_{CO}$, i.e. $E_d = E_{d0} - \varepsilon_1\Theta_{CO}$. Obviously (b) has to be preferred since there is no experimental evidence (in the considered coverage range) that either the CO sticking probability or the frequency factor for desorption does depend markedly on $\Theta_{CO}$, and because, as demonstrated in Chap. 4, the CO molecules are repelling each other. The plotted curves lead to a value $\varepsilon_1 = 16.4$ kcal/mole. The Arrhenius type plot $\ln\{[p_{CO}(1 - \Theta_{CO})/\Theta_{CO}]T_s^{-3}\}\big|_{\Theta_{CO}=0}$ vs $T_s^{-1}$ shown in Fig. 8.10 and obtained from the straight lines in Fig. 8.9 extrapolated to $\Theta_{CO} = 0$, yields $K = 12.5$ mbar K$^{-3}$ and $E_{d0} = 31.7$ kcal/mole. With these values of the parameters $K$, $E_{d0}$ and $\varepsilon_1$ and by using the calibration function, the "experimental" isotherms for the $T_s$ values used in the measurement were calculated. They are plotted as solid lines in Fig. 8.8.

The second example concerns the adsorption of hydrogen on the stepped Pt(997) surface [Poelsema, Mechtersheimer, Comsa, 1981a; 1981b]. This example is given here in order to show that information on coverage can be obtained by monitoring not only the specular beam but also any other diffracted beam. It will be also shown that due to the particular scattering geometry one may distinguish between adsorbates located on terraces and on step sites, respectively. Because the example here concerns H adatoms, which are not purely diffuse scatterers, only a qualitative discussion will be given, in order to avoid too complicated arguments. The Pt(997) surface consists of (111) terraces, about 20 Å wide, separated by monatomic steps along a <110> direction. The diffraction patterns obtained at $p_{H_2} = 2 \times 10^{-11}$ mbar and $8.5 \times 10^{-8}$ mbar at $T_s = 332$ K with a glancing He beam ($\vartheta_i = 85°$) in the stepdown direction are shown in Fig. 8.11 as solid and dashed curves, respectively. The positions of the diffraction peaks correspond to the Bragg formula (6.5) with the parameters of the Pt(997) surface. It has been demonstrated that the higher order diffraction peaks stem preferentially from scattering contributions from the terraces, while the zeroth order peak originates in He scattering from the ledges [Poelsema, Mechtersheimer, Comsa, 1981a]. A comparison of the curves shows that the attenuation of the higher order peaks is substantially smaller than that of the zero order one, showing that, as expected,

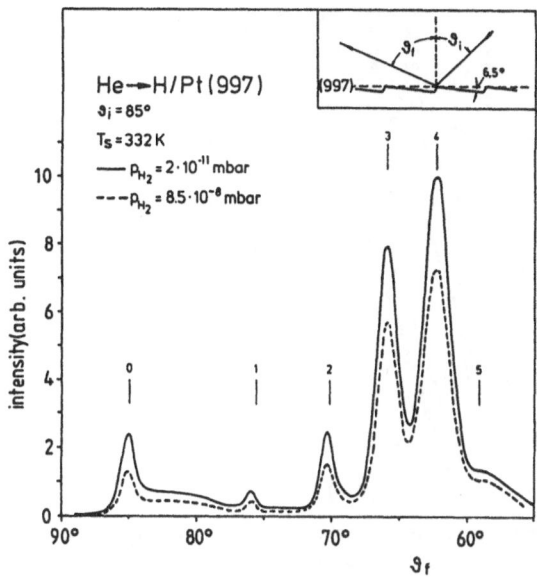

**Fig. 8.11.** He diffraction patterns obtained at $T_s = 332\,\mathrm{K}$ from a clean Pt(997) surface (*solid*) and in the presence of a $8.5 \times 10^{-8}$ mbar equilibrium $H_2$ pressure. The 63 meV He beam was incident in the step-down direction (along the $[7, 7, -18]$ azimuth) at $\vartheta_i = 85°$ with respect to the normal to the macroscopic plane of the Pt(997) surface. The vertical lines indicate the peak positions according to (6.5)

the H coverage on the terraces is less than on the step sites. It is thus possible to distinguish between adsorption on terrace sites and on step sites. From patterns like these, measured at various equilibrium $H_2$ pressures and at a constant temperature, isotherms have been constructed for each of the diffraction orders. Repeating the procedure at several temperatures a family of isotherms is obtained for each diffraction order. These were transformed into isosteres and then through the Clausius–Clapeyron equation an isosteric heat of adsorption could be associated with each diffraction order. It appears that the higher order peaks yield a unique value for the isosteric heat of adsorption 18 kcal/mole (extrapolated to $\Theta_H = 0$), while the zero order peak yields a value of 21 kcal/mole. In view of the specific contribution of the scattering from terraces and ledges to higher order peaks and, respectively, to the zero order peak mentioned above, the 18 and 21 kcal/mole heats of adsorption are associated with terrace sites and step sites, respectively.

### 8.2.3  Isobars

In order to investigate the 2D-condensation of Xe on Pt(111) in the higher temperature range, it is necessary to work at higher 3D-Xe pressures. Thus, as noted in Sect. 8.2.1 it is more appropriate to monitor isobars than isotherms. A family of isobars is shown in Fig. 8.12. For $T_s \leq 120\,\mathrm{K}$, i.e. below the critical temperature, each isobar contains a segment with very large slope (below the dash-dotted line). Within this segment a small decrease in temperature results in a dramatic decrease of $I/I_0$, i.e. in a dramatic coverage increase. This is characteristic for the coexistence of a 2D-gas phase with a 2D-condensed phase. The isobars above $T_s \geq 120\,\mathrm{K}$ and the isobar segments above the dash-dotted line indicate the presence of a unique 2D-fluid (gas) phase.

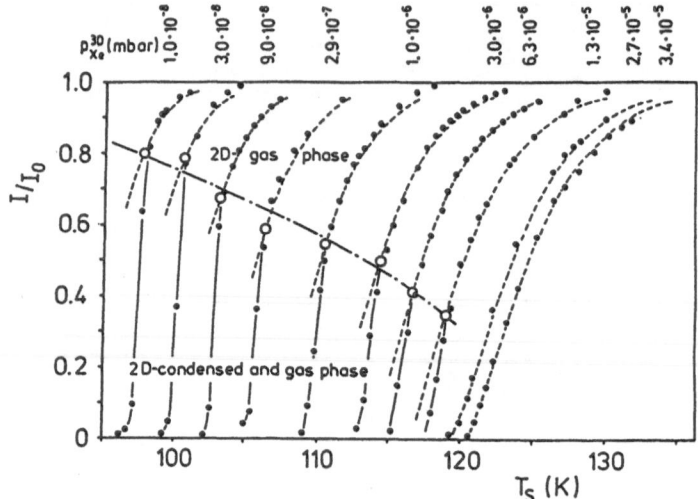

**Fig. 8.12.** Isobars, i.e. the relative He specular peak height versus the Pt(111) surface temperature $T_s$ at constant 3D-Xe equilibrium pressure, $p_{Xe}$. The dashed curves are calculated isobars, corresponding to the 2D-Xe gas phase. The dash-dotted curve connects the open circles designating the 2D-gas-condensed phase transition (see text)

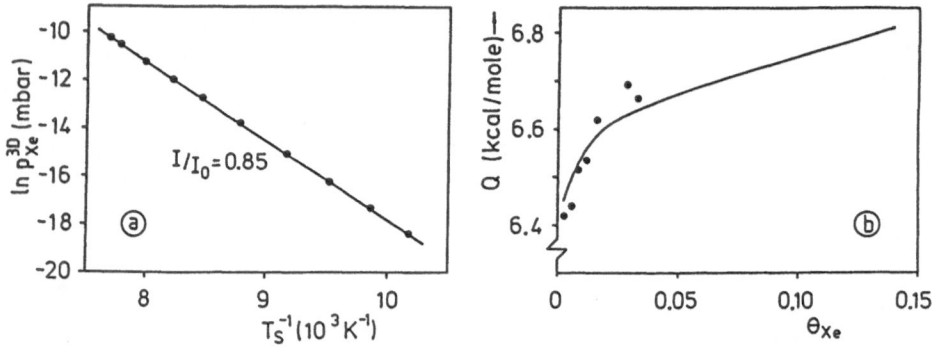

**Fig. 8.13.** (a) Isostere, i.e. 3D-Xe pressure versus the reciprocal Pt(111) surface temperature at constant Xe coverage, $\Theta_{Xe} = 0.009$ ($I/I_0 = 0.85$). (b) Isosteric heat of adsorption for the 2D-Xe gas phase as a function of Xe coverage

The isosteric heat of adsorption for the region with a unique 2D-gas phase is obtained by transforming the isobars into isosteres. We cut the isobars along a horizontal line $I/I_0$ = constant $\rightarrow \Theta_{Xe}$ = constant and determine for each intersection with the isobars the couple of values $p_{Xe}^{3D}$ and $T_s$. These values are then plotted in a Clausius–Clapeyron $\ln p_{Xe}^{3D}$ vs $T_s^{-1}$ plot as shown in Fig. 8.13a for $I/I_0 = 0.85$ ($\Theta_{Xe} = 0.009$). From the slope a heat of adsorption of 6.51 kcal/mole is obtained. By repeating the procedure for various $I/I_0$ = constant values ($\Theta_{Xe}$ = constant) the dependence of the isosteric heat of adsorption on $\Theta_{Xe}$ shown in Fig. 8.13b is obtained. The obvious increase of the heat of adsorption with $\Theta_{Xe}$ is consistent with the attractive Xe-Xe interaction leading to the 2D-

Xe condensation (island formation) observed both here ("vertical" segments in Fig. 8.12) and in Figs. 4.6 – 8.

The same procedure can be applied to extract a heat of adsorption by intersecting the "vertical" segments, see e.g. [Hill, 1949]. The corresponding Clausius–Clapeyron plot obtained for $I/I_0 = 0.3$ is shown in Fig. 8.14 and yields a heat of adsorption of $7.75 \pm 0.10$ kcal/mole. This value does not vary significantly with $\Theta_{Xe}$. It represents the difference between the enthalpy of the 3D-Xe gas and that of the 2D-Xe condensed phase (Xe islands). This quantity should be in fact approximately equal to the sum of the heat of adsorption from the 3D-Xe gas phase into the 2D-gas phase at $\Theta_{Xe} \approx 0$ (6.4 kcal/mole, Fig. 8.13b) and the 2D-latent heat of vaporization from the 2D-condensed phase into the 2D-gas phase (1.1±0.1 kcal/mole, Sect. 4.5) [Poelsema, Verheij, Comsa, 1985b]. The agreement is remarkable.

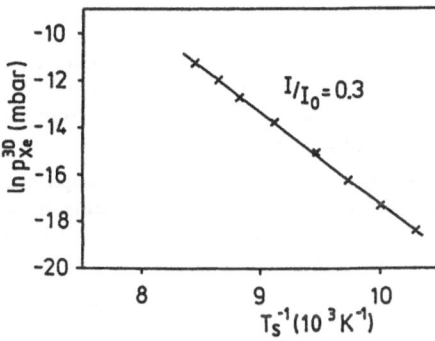

**Fig. 8.14.** Semilog plot of the 3D-Xe pressure required for 2D condensation versus the reciprocal Pt(111) surface temperature at a 2D-Xe coverage corresponding to a He reflectivity of $I/I_0 = 0.3$

## 8.2.4 Kinetics of Desorption

The possibility of continuously monitoring the surface coverage gives the opportunity of investigating in detail the desorption process as well. The two examples below correspond to the usual procedures. First, the adsorbed layer is built up at low temperatures; then alternatively the surface is heated rapidly up to some temperature $T_s$ which is maintained constant or the surface temperature is increased linearly (temperature programmed desorption — TPD). The most widespread way to follow the desorption is to monitor the evolution of the ambient pressure; under appropriate conditions the monitored pressure is proportional to the rate of desorption. The reflectivity of the He beam which is monitored here supplies the instantaneous surface coverage, i.e. the amount of adsorbate which has not yet been desorbed. (The same type of desorption transients are obviously obtained when the coverage is monitored by other means like Auger-electron-spectroscopy or work-function measurements.)

The desorption transients presented in Figs. 8.15 and 16 are isothermal desorption ($T_s$ = constant) and TPD curves, respectively. In both cases, at the end of the desorption process the relative reflectivity becomes $I/I_0 = 1$. This demonstrates that the CO adsorption on Pt(111) is fully reversible. The solid curves are obtained by inserting in the molecular desorption equation the pa-

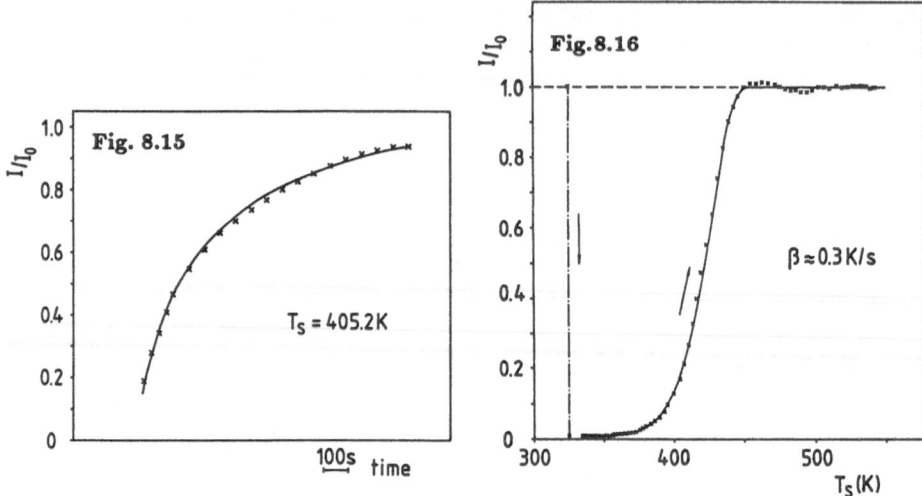

**Fig. 8.15.** Isothermal desorption curve, i.e. the height of the specular He peak versus time at constant surface temperature $T_s = 405.2\,K$ for CO/Pt(111). The solid curve represents the behaviour calculated with parameters deduced from the adsorption kinetics and from the equilibrium data (Figs. 4.2 and 8.8), respectively. *No* free parameter has been used

**Fig. 8.16.** Temperature programmed desorption of CO from Pt(111); i.e. height of the specular He peak versus surface temperature (corrected for Debye–Waller effects) during heating of the surface (heating rate $\beta = 0.3\,Ks^{-1}$). The surface has been initially saturated with CO at $T_s = 325\,K$ (*vertical dash-dotted line*). The solid curve shows the calculated behaviour (see also Fig. 8.15)

rameters $E_{d0}$, $\varepsilon_1$ and $K$ determined in the equilibrium experiments described in Sect. 8.2.2. (The preexponential desorption factor $\nu$ is obtained from $K$ and the value of the CO-sticking probability on Pt(111) well-known from the literature.) Note that there is no free fitting parameter; in spite of this, the fit of the data is remarkable. This is a consequence of the accurate and non-damaging coverage monitoring by means of thermal He scattering. It demonstrates also that — at least for CO/Pt(111) — the desorption process which can be monitored only under non-equilibrium conditions ($p_{3D}^{CO} \approx 0$) is still governed by the "quasi"-equilibrium parameters.

## 8.3 Measurement of Relative Changes of the Sticking Probability

The sticking probability is a central quantity in surface physics. Unfortunately, the accurate determination of absolute values of the sticking probability is still a challenge. As already discussed in Chap. 3, TEAS is not an adequate method to determine absolute sticking probabilities; on the contrary, such values are needed to get accurate cross-sections for diffuse scattering, $\Sigma$. However, due to the large value of $\Sigma$ for most adsorbates, TEAS can be used to accurately determine changes in the sticking probability as a function of any parameter also without the knowledge of the absolute value of the sticking probability. The only condition for such a measurement is that the effective value of $\Sigma$ is not influenced by changes of the parameters [Poelsema, Verheij, Comsa, 1984b]. This latter fact can be directly checked by He reflectivity measurements at constant coverage. The examples given below refer to changes in sticking probability with surface temperature; such values are needed for the understanding of the adsorption process.

Figure 8.17 shows two hydrogen adsorption curves on Pt(111), taken at different surface temperatures $T_s = 90\,\mathrm{K}$ (×) and $T_s = 240\,\mathrm{K}$ (•). The curves coincide after contracting the 90 K exposure scale by a factor of 1.30. Thus the sticking probability of hydrogen at $T_s = 240\,\mathrm{K}$ is 30% *larger* than at $T_s = 90\,\mathrm{K}$ [Poelsema, Verheij, Comsa, 1985a]. It has been checked, by measuring the relative He specular-peak height at constant hydrogen coverage as a function of temperature, that the nature of the lateral distribution of hydrogen does not change in this temperature range. Since there is no other reason for a change of the effective scattering cross-section [Poelsema, Verheij, Comsa, 1984b], it can be concluded, that $\Sigma$ is constant in this temperature range. Similar measurements for CO/Pt(111) show that the sticking probability of CO *decreases* by 14% between $T_s = 90\,\mathrm{K}$ and 300 K [Poelsema, Verheij, Comsa, 1985a].

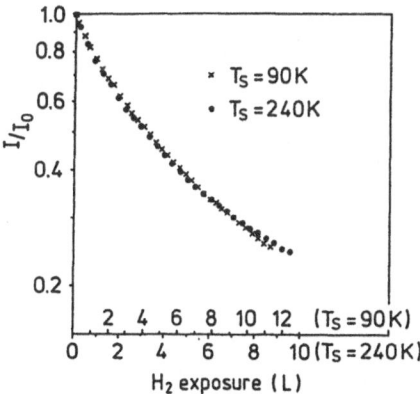

Fig. 8.17. Adsorption curves, i.e. the relative He specular peak height versus $H_2$ exposure, of Pt(111) at constant temperature: $T_s = 90\,\mathrm{K}$ (×) and $T_s = 240\,\mathrm{K}$ (•), respectively. The ratio of the exposure scales has been determined by the coincidence condition of the curves (see text)

Elliot, Jónsson, Miller, Weare [1985] have followed the same train of thought to estimate the sticking probability of CO on Au(111). After having established that the angular dependence of the He specular peak height reduction at a fixed (low) CO coverage on Au(111) closely resembles that of CO on Pt(111), they concluded that the size of the cross-section for diffuse He scattering from CO adsorbates on Au(111) and on Pt(111) is similar. From their CO adsorption curves on Au(111) they could then infer that the CO sticking probability on Au(111) amounts to about 0.001.

# References

Armand, G., D. Gorse, J. Lapujoulade, J.R. Manson [1987]: Europhys. Lett. **3**, 1113 (Sect. 7.3)
Armand, G., J. Lapujoulade, Y. Lejay [1977]: Surface Sci. **63**, 143 (Sect. 7.3)
Armand, G., J.R. Manson [1979]: Phys. Rev. Lett. **43**, 1839 (Sect. 3.1)
Avron, J.E., L.S. Balfour, C.G. Küper, J. Landau, S.G. Lipson, L.S. Schulman [1980]: Phys. Rev. Lett. **45**, 814 (Sect. 3.1)

Bagus, P.S., K. Hermann [1986]: Phys. Rev. B, **33**, 2987 (Sect. 3.3)
Barker, J.A., D. Henderson, F.F. Abraham [1981]: Physica **106A**, 226 (Sect. 4.5)
Becker, E.W., K. Bier [1954]: Z. f. Naturforsch. **99**, 975 (Chap. 1, Sect. 2.1)
Beeby, J. [1971]: J. Phys. C, **4**, L359 (Sects. 3.1, 3)
Bernasek, S.L., K. Lenz, B. Poelsema, G. Comsa [1988]: Surface Sci. **183**, L319 (Sect. 5.2.2)
Bernasek, S.L., G.A. Somorjai [1975]: J. Chem. Phys. **62**, 3149 (Sect. 8.1.2)
Besocke, K., B. Krahl–Urban, H. Wagner [1977]: Surface Sci. **68**, 39 (Sect. 6.1)
Binnig, G., H. Rohrer, Ch. Gerber, E. Weibel [1982]: Phys. Rev. Lett. **49**, 57 (Chap. 1)
Boato, G., P. Cantini, L. Mattera [1976]: Surface Sci. **55**, 141 (Chap. 1, Sect. 2.1)
Boato, G., P. Cantini, R. Tatarek [1976]: J. Phys. F, **6**, L237 (sect. 3.1)
Bonzel, H.P. [1983]: in *Surface Mobilities on Solid Materials*, ed. by Vu Thien Binh (Plenum, New York) and references therein (Sect. 6.1)
Bosanac, S.D., M. Sunjic [1985]: Chem. Phys. Lett. **115**, 75 (Sects. 3.3, 4)
Bruch, L.W. [1983]: private communication (Sect. 4.5)
Brusdeylins, G., R.B. Doak, J.P. Toennies [1980]: Phys. Rev. Lett. **44**, 1417 (Chap. 1)
Burton, W.K., N. Cabrera, F.C. Frank [1951]: Phil. Trans. Roy. Soc. London **243A**, 299 (Sect. 7.1)
Butz, H.P., R. Feltgen, H. Pauly, H. Vehmeyer [1971]: Z. Physik **247**, 70 (Sect. 3.1)

Campbell, C.T., G. Ertl, H. Kuipers, J. Segner [1981]: Surface Sci. **107**, 207 (Sect. 3.3)
Cardillo, M.J., G.I. Becker [1978]: Phys. Rev. Lett. **40**, 1148 (Chap. 1, Sect. 2.1)
Celli, V. [1985]: private communication (Sect. 3.5)
Celli, V., N. Garcia, J. Hutchison [1979]: Surface Sci. **87**, 112 (Sect. 3.1)
Cherns, D. [1977]: Phil. Mag. **36**, 1429 (Sects. 6.1, 6.5.2)
Christmann, K., G. Ertl [1976]: Surface Sci. **60**, 635 (Sect. 8.1.2)
Christmann, K., G. Ertl, T. Pignet [1976]: Surface Sci. **54**, 365 (Sect. 8.1.2)
Collins, D.M., W.E. Spicer [1977a]: Surface Sci. **69**, 85 (Sect. 8.1.2)
Collins, D.M., W.E. Spicer [1977b]: Surface Sci. **69**, 114 (Sect. 8.1.2)
Comsa, G. [1979]: Surface Sci. **81**, 57 (Sect. 2.1)
Comsa, G., B. Poelsema [1985]: Appl. Phys. A, **38**, 153 (Sects. 3.5, 6.5)
Comsa, G., G. Mechtersheimer, B. Poelsema [1979]: Surface Sci. **89**, 123 (Sect. 6.2)
Conrad, E.H., L.R. Allen, D.L. Blanchard, T. Engel [1987]: Surface Sci. **187**, 265 (Sects. 7.1, 7.2.1, 7.2.2)
Conrad, E.H., L.R. Allen, D.L. Blanchard, T. Engel [1988]: Surface Sci. **198**, 207 (Sect. 7.2.2)
Conrad, E.H., R.M. Aten, D.S. Kaufman, L.R. Allen, T. Engel [1986a]: J. Chem. Phys. **84**, 1015 (Sects. 6.3, 7.1, 7.2, 7.2.1, 7.2.2)
Conrad, E.H., R.M. Aten, D.S. Kaufman, L.R. Allen, T. Engel, M. Den Nijs, E.K. Riedel [1986b]: J. Chem. Phys. **85**, 4756 (Sects. 7.1, 7.2, 7.2.2)
Cowley, J.M., H. Shuman [1973]: Surface Sci. **38**, 53 (Sect. 6.1)
Crossley, A., D.A. King [1980]: Surface Sci. **95**, 131 (Sect. 5.1)

David, R., K. Kern, P. Zeppenfeld, G. Comsa [1986]: Rev. Sci. Instrum. **57**, 2771 (Sect. 2.2)

De Miguel, J.J., A. Sanchez, A. Cebollada, J.M. Gallego, J. Ferron, S. Ferrer [1987]: Surface Sci. **189/190**, 1062 (Sect. 6.6)

Den Nijs, M., E.K. Riedel, E.H. Conrad, T. Engel [1985]: Phys. Rev. Lett. **55**, 1989 (Sects. 7.1, 7.2)

Den Nijs, M., E.K. Riedel, E.H. Conrad, T. Engel [1986]: Phys. Rev. Lett. **57**, 1279 (Sects. 7.1, 7.2, 7.2.2)

DiFoggio, R., R. Gomer [1982]: Phys. Rev. B, **25**, 3490 (Sects. 6.1, 5.2.2)

Dobson, P.J., B.A. Joyce, J.H. Neave, J. Zhang [1987]: J. of Cryst. Growth, **81**, 1 (Sect. 6.6)

Elliot, F., H. Jónsson, D.R. Miller, J. Weare [1985]: J. Vac. Sci. Technol. A, **3**, 1965 (Sect. 8.3)

Engel, T. [1078]: J. Chem. Phys **69**, 373 (Sect. 3)

Engel, T., H. Kuipers [1979]: Surface Sci. **90**, 162 (Sect. 4.2)

Engel, T., K.H. Rieder [1982]: *Structural Studies of Surfaces wtih Atomic and Molecular Beam Diffraction*, Springer Tracts in Modern Physics **91** (Springer, Heidelberg, Berlin) (Chap. 1, Sect. 8.1.1)

Esbjerg, N., J.K. Norskov [1980]: Phys. Rev. Lett. **45**, 807 (Sect. 3.1)

Estermann, J., O. Stern [1930]: Z. Physik **61**, 95 (Sect. 3.1)

Fabre, F., D. Gorse, B. Salanon, J. Lapujoulade [1987a]: Europhys. Lett. **3**, 737 (Sects. 7.1, 7.2.2)

Fabre, F., D. Gorse, B. Salanon, J. Lapujoulade [1987b]: Physique **48**, 1017 (Sects. 6.3, 7.1, 7.2.1, 7.2.2)

Fabre, F., B. Salanon, J. Lapujoulade [1988a]: *The structure of surfaces II*, ed. by van der Veen, J.F., M.A. van Hove (Springer, Berlin) p.520 (Sects. 7.1, 7.2.2)

Fabre, F., B. Salanon, J. Lapujoulade [1988b]: Solid State Commun., accepted (Sects. 7.1, 7.2.2)

Feenstra, R.M. [1988]: *Proc. Second Int. Conf. on Scanning Tunneling Microscopy*, J. Vac. Sci. Technol. A, **6**, 249ff (Chap. 1)

Frankl, D.R., D. Wesner, S.V. Krishnaswamy, G. Derry, T.O. Gorman [1978]: Phys. Rev. Lett. **41**, 60 (Sect. 3.1)

Frenken, J.W.M., F. Huussen, J.F. van der Veen [1987]: Phys. Rev. Lett. **58**, 401 (Sect. 7.3)

Garcia, N. [1976]: Phys. Rev. Lett. **37**, 912 (Sect. 3.1)

Garcia, N. [1987]: *Proc. First Int. Conf. on Scanning Tunneling Microscopy*, Surface Sci. **181** (Chap. 1)

Garcia, N., V. Celli, F.O. Goodman [1979]: Phys. Rev. B, **19**, 634 (Sect. 3.1)

Garcia, N., J. Ibañez, J. Solana, N. Cabrera [1976]: Solid State Commun. **20**, 1559 (Sect. 3.1)

Glachant, A., U. Bardi [1979]: Surface Sci. **87**, 187 (Sect. 5.2.1)

Gomez, L.J., S. Bourgeal, J. Ibañez, M. Salmeron [1985]: Phys. Rev. B, **31**, 2551 (Sect. 6.6)

Gorse, D., J. Lapujoulade [1985]: Surface Sci. **162**, 847 (Sect. 7.2.2)

Greenler, R.G., K.D. Bruch, K. Kretzschmar, R. Klauser, A. Bradshaw, B.F. Hayden [1985]: Surface Sci. **152/153**, 338 (Sect. 5.1)

Gumhalter, B., W.K. Liu [1984a]: Surface Sci. **148**, 371 (Sects. 3.3, 3.4)

Gumhalter, B., W.K. Liu [1984b]: Surface Sci. **157**, 539 (Sects. 3.3, 3.4)

Hamann, D. [1981]: Phys. Rev. Lett. **46**, 1227 (Sect. 3.1)

Harris, J., A. Liebsch [1982]: Phys. Rev. Lett. **49**, 341 (Sect. 3.1)

Henzler, M. [1970]: Surface Sci. **22**, 12 (Sect. 6.1)

Henzler, M. [1978]: Surface Sci. **73**, 240 (Sect. 6.1)

Henzler, M. [1982]: Appl. Surface Sci. **11/12**, 450 (Sect. 6.1)

Hill, T. [1949]: J. Chem. Phys. **17**, 520 (Sect. 8.2.3)

Horne, J., D.R. Miller [1976]: J. Vac. Sci. Technol. **13**, 351 (Chap. 1, Sect. 2.1)

Horne, J., S.C. Yerkes, D.R. Miller [1980]: Surface Sci. **93**, 47 (Sect. 3.1)

Houston, J.E., R.L. Park [1970]: Surface Sci. **21**, 209 (Chap. 1)

Houston, J.E., R.L. Park [1971]: Surface Sci. **26**, 269 (Chap. 1)

Ibach, H., W. Erley, H. Wagner [1980]: Surface Sci. **92**, 29 (Sect. 8.2.2)

Ibañez, J., N. Garcia, J.M. Rojo [1983]: Phys. Rev. B, **28**, 3174 (Sect. 3.3)

Ibañez, J., N. Garcia, J.M. Rojo, N. Cabrera [1982]: Surface Sci. **117**, 23 (Sect. 3.3)

Jayanthi, C.S., G. Armand, J.R. Manson [1985]: Surface Sci. **154**, L247 (Sects. 7.2, 3)

Jónsson, H. [1986]: Ph. D. Thesis, University of San Diego (USA) (Sects. 3.3, 4)

Jónsson, H., J. Weare, A.C. Levi [1984a]: Phys. Rev. B, **30**, 2441 (Sects. 3.3, 4)

Jónsson, H., J. Weare, A.C. Levi [1984b]: Surface Sci. **148**, 126 (Sects. 3.3, 4)

Joyce, B.A., P.J. Dobson, J.H. Neave, J. Zhang [1986]: Surface Sci. **178**, 110 (Sect. 6.6)

Kantrowitz, A., J. Grey [1951]: Rev. Sci. Instrum. **22**, 328 (Chap. 1., Sect. 2.1)

Kawamura, T., P.A. Maksym [1985]: Surface Sci. **161**, 12 (Sect. 6.6)

Kern, K., R. David, R.L. Palmer, G. Comsa [1986]: Phys. Rev. Lett. **56**, 620 (Sect. 4.5)

Kosterlitz, J.M. [1974]: J. Phys. C, **7**, 1046 (Sect. 7.1)

Kosterlitz, J.M., D.J. Thouless [1973]: J. Phys. C, **6**, 1181 (Sect. 7.1)

Krebs, H.J., H. Lüth [1977]: Appl. Phys. **14**, 337 (Sect. 3.3)

Lahee, A.M., J.R. Manson, J.P. Toennies, C. Wöll [1986]: Phys. Rev. Lett. **57**, 471 and ibid, 2331 (Sects. 6.1, 6.5.3)

Lapujoulade, J. [1981]: Surface Sci. **108**, 526 (Sects. 6.1, 3)

Lapujoulade, J. [1986]: Surface Sci. **178**, 406 (Sect. 7.1)

Lapujoulade, J., J. Perreau, A. Kara [1983]: Surface Sci. **129**, 59 (Sect. 7.2)

Lee, J., J.P. Cowin, L. Wharton [1983]: Surface Sci. **130**, 1 (Sects. 4.2, 8.1.1)

Lent, C.S., P.I. Cohen [1984]: Surface Sci. **139**, 121 (Sect. 6.6)

Lenz, K. [1987]: Ph. D. Thesis, University of Bonn (F.R.G.), Jül-2141 (Sect. 8.1.2)

Lenz, K., B. Poelsema, S.L. Bernasek, G. Comsa [1987]: Surface Sci. **189/190**, 431 (Sect. 5.2.2)

Levi, A.C., R. Spadacini, G.E. Tommei [1982]: Surface Sci. **121**, 504 (Sect. 6.1)

Levi, A.C., H. Suhl [1979]: Surface Sci. **88**, 221 (Sect. 7.3)

Liang, K.S., E.B. Sirota, K.L. D'Amico, G.J. Hughes, S.K. Sinha [1987]: Phys. Rev. Lett. **59**, 2447 (Sect. 7.1)

Liebsch, A., J. Harris, B. Salanon, J. Lapujoulade [1982]: Surface Sci. **123**, 338 (Sect. 3.1)

Lin, T.H., G.A. Somorjai [1981]: Surface Sci. **107**, 573 (Sect. 3.3)

Linke, U. [1988]: private communication (Sect. 6.1)

Linke, U., Poelsema [1985]: J. Phys. E: Sci. Instrum. **18**, 26 (Sects. 2.3, 5.1)

Liu, W.K., B. Gumhalter [1986]: *Report IC-7*, International Center for Theoretical Physics, Trieste (Sects. 3.3, 4)

Lu, T.M., M.G. Lagally [1982]: Surface Sci. **120**, 47 (Sect. 6.1)

Manson, J.R. [1988]: private communication (Sect. 3.5)

Mason, B.F., B.R. Williams [1972]: Rev. Sci. Instrum. **43**, 375 (Chap. 1)

Mochrie, S.G.J. [1987]: Phys. Rev. Lett. **59**, 304 (Sects. 7.1, 7.2.1, 7.2.2)

Mullins, W.M. [1959]: J. Appl. Phys. **30**, 77 (Sect. 6.1)

Neave, J.H., B.A. Joyce, P.J. Dobson, N. Norton [1983]: Appl. Phys. A, **31**, 1 (Sect. 6.6)

Osakabe, N., Y. Tanishiro, K. Yagi, G. Honjo [1981]: Surface Sci. **102**, 424 (Sect. 6.1)

Pauly, H., J.P. Toennies [1956]: *The Study of Intermolecular Potentials with Molecular Beams at Thermal Energies*, Advances in atomic and molecular physics, Vol. 1 (Academic, New York) (Sect. 2.1)

Phillips, J.M., L.W. Bruch, R.D. Murphy [1981]: J. Chem. Phys. **75**, 5097 (Sect. 4.5)

Poelsema, B. [1983]: unpublished results (Sect. 3.3)

Poelsema, B., L.S. Brown, K. Lenz, L.K. Verheij, G. Comsa [1986]: Surface Sci. **171**, L395 (Sects. 4.2, 8.1.1)

Poelsema, B., G. Comsa [1985a]: Comments At. Mol. Phys. **17**, 37 (Sect. 3.5)

Poelsema, B., G. Comsa [1985b]: Faraday, Discuss. Chem. Soc. **80**, 115 (Sects. 3.5, 6.5)

Poelsema, B., S.T. de Zwart, G. Comsa [1982]: Phys. Rev. Lett. **49**, 578 (Chap. 1, Sects. 3, 3.3, 3.5.1, 4.3, 5.1)

Poelsema, B., S.T. de Zwart, G. Comsa [1983]: Phys. Rev. Lett. **51**, 522 (E) (Chap. 1, Sects. 3, 3.3, 3.5.1, 4.3, 5.1)

Poelsema, B., K. Lenz, L.S. Brown, L.K. Verheij, G. Comsa [1985]: Surface Sci. **162**, 1011 (Sects. 4.4, 5.1, 6.5)

Poelsema, B., G. Mechtersheimer, G. Comsa [1981a]: Surface Sci. **111**, 519 (Sects. 2.2, 8.2.1)

Poelsema, B., G. Mechtersheimer, G. Comsa [1981b]: Surface Sci. **111**, L728 (Sects. 8.2, 8.2.1)

Poelsema, B., R.L. Palmer, G. Comsa [1982]: Surface Sci. **123**, 152 (Sect. 8.2.2)

Poelsema, B., R.L. Palmer, S.T. De Zwart, G. Comsa [1983]: Surface Sci. **126**, 641 (Sects. 3.3, 3.4)

Poelsema, B., R.L. Palmer, G. Mechtersheimer, G. Comsa [1982]: Surface Sci. **117**, 50 (Sects. 2.3, 3.3, 6.3)

Poelsema, B., L.K. Verheij, G. Comsa [1982]: Phys. Rev. Lett. **49**, 1731 (Sect. 5.1)

Poelsema, B., L.K. Verheij, G. Comsa [1983a]: Phys. Rev. Lett. **51**, 2410 (Sects. 4.5, 8.2.3)

Poelsema, B., L.K. Verheij, G. Comsa [1983b]: *Proc. Symp. Surface Sci.* (Obertraun), ed. by P. Braun and G. Betz, p. 174 (Sect. 5.1)

Poelsema, B., L.K. Verheij, G. Comsa [1984a]: Phys. Rev. Lett. **53**, 2500 (Sects. 4.4, 6.5.1)

Poelsema, B., L.K. Verheij, G. Comsa [1984b]: Surface Sci. **148**, 117 (Sects. 8.2, 8.3)

Poelsema, B., L.K. Verheij, G. Comsa [1985a]: Surface Sci. **152/153**, 496 (Sect. 8.3)

Poelsema, B., L.K. Verheij, G. Comsa [1985b]: Surface Sci. **152/153**, 851 (Sects. 4.5, 8.2.3)

Reutt–Robey, J.E., D.J. Doren, Y.J. Chabal, S.B. Christman [1988]: Phys. Rev. Lett. **61**, 2778 (Sect. 5.1)

Rieder, K.H. [1982]: Surface Sci. **117**, 13 (Sect. 3.1)

Roosendaal, H. [1981]: in *Sputtering by Particle Bombardment I* (Chap. 5), ed. by R. Behrisch (Springer, Berlin, Heidelberg, New York) (Sects. 4.4, 6.5.1)

Salanon, B., F. Fabre, J. Lapujoulade, W. Selke [1988]: Surface Sci., to be published (Sect. 7.2.2)

Salméron, M., R.J. Gale, G.A. Somorjai [1977]: J. Chem. Phys. **67**, 5324 (Sect. 8.1.2)

Salméron, M., R.J. Gale, G.A. Somorjai [1979]: J. Chem. Phys. **70**, 2807 (Sect. 8.1.2)

Sánchez, A., S. Ferrer [1987]: Surface Sci. **187**, L587 (Sect. 6.4)

Sánchez, A., J. Ibañez, R. Miranda, S. Ferrer [1986]: Surface Sci. **178**, 917 (Chap. 5)

Schulze, G., M. Henzler [1978]: Surface Sci. **73**, 553 (Sect. 6.1)

Smith, D.L., R.P. Merrill [1970]: J. Chem. Phys. **52**, 5861 (Chap. 1, Sect. 3)

Sommerfeld, A. [1959]: *Vorlesungen über theoretische Physik, Band 4: Optik*, Chap. 5 (Geest and Portig, Leipzig) (Sect. 6.2)

Somorjai, G.A. [1981]: *Chemistry in Two Dimensions; Surfaces* (Cornell University Press, Ithaca, New York (USA)) (Sect. 6.1)

Spadacini, R., G.E. Tommei [1983]: Surface Sci. **133**, 216 (Sect. 6.1)

Steininger, H. [1982]: Ph. D. Thesis, University of Aachen (F.R.G.), Jül-1763 (Sect. 5.2.2)

Steininger, H., S. Lehwald, H. Ibach [1982]: Surface Sci. **123**, 264 (Sects. 3.3, 4.2)

Stern, O. [1929]: Naturwissenschaften **17**, 391 (Chap. 1)

Tenner, M.G., M.E.M. Spruit, E.W. Kuipers, A.W. Kleijn [1987]: Surface Sci. **189/190**, 656 (Sect. 6.3)

Toennies, J.P. [1984]: J. Vac. Sci. Technol. A, **2**, 1055 and references therein (Chap. 1)

Van Hove, J.M., C.S. Lent, P.R. Pukite, P.I. Cohen [1983]: J. Vac. Sci. Technol. B, **1**, 741 (Sect. 6.6)

Vehmeyer, H., R. Feltgen, P. Chakraborti, M. Düker, F. Torell, H. Pauly [1976]: Chem. Phys. Lett. **42**, 507 (Sect. 3.5)

Verheij, L.K. [1982]: Surface Sci. **114**, 667 (Sects. 6.1, 6.5.2)

Verheij, L.K., J. Lux, A.B. Anton, B. Poelsema, G. Comsa [1987]: Surface Sci. **182**, 390 (Sect. 8.2.2)

Verheij, L.K., J. Lux, B. Poelsema [1984]: Surface Sci. **144**, 385 (Sects. 6.3, 4)

Verheij, L.K., J. Lux, B. Poelsema, G. Comsa [1987]: Surface Sci. **187**, L581 (Sect. 2.1)

Verheij, L.K., B. Poelsema, G. Comsa [1985]: Surface Sci. **162**, 858 (Sects. 5.1, 6.1, 3, 4)

Verheij, L.K., J.A. Van den Berg, D.G. Armour [1982]: Surface Sci. **122**, 216 (Sects. 6.1, 6.5.2)

Villain, J., D.R. Grempel, J. Lapujoulade [1985]: J. Phys. F, **15**, 809 (Sects. 7.1, 7.2.2)

Vlieg, E., A.W. Denier van der Gon, J.F. Van der Veen, J.E. MacDonald, C. Norris [1988]: Phys. Rev. Lett. **61**, 2241 (Sect. 6.6)

Wachs, I.E., R.J. Madix [1976]: Surface Sci. **58**, 590 (Sect. 8.1.2)

Weeks, J.D. [1980]: *Ordering in Strongly Fluctuating Condensed Matter Systems*, ed. by T. Riste (Plenum, New York) (Sect. 7.1)

Welkie, D.G., M.G. Lagally [1979]: J. Vac. Sci. Technol. **16**, 784 (Sect. 6.1)

Wilsch, H., K.H. Rieder [1983]: J. Chem. Phys. **78**, 7491 (Sect. 3.3)

Yinnon, A.T., R. Kosloff, R.B. Gerber, B. Poelsema, G. Comsa [1988]: J. Chem. Phys. **88**, 3722 (Sects. 3.3, 3.5, 4.4)

Zaremba, E. [1985]: Surface Sci. **151**, 91 (Sect. 4.4)

Zaremba, E., W. Kohn [1976]: Phys. Rev. B, **13**, 2270 (Sect. 3.1)

Zeppenfeld, P. [1988]: private communication (Sect. 7.2.1)

Zeppenfeld, P., K. Kern, R. David, G. Comsa [1989]: Phys. Rev. Lett. **62**, 63 (Sects. 7.1, 7.2.1, 7.2.2, 7.3)

# Subject Index

Adsorbate
— perfectly diffuse scatterer  14, 23ff, 29ff, 47, 83f, 91
— with finite He reflectivity  22f, 26ff, 32, 47, 83, 84ff
Adsorption
— coverage  15, 22ff, 29ff, 83ff
— curve  14, 22, 30, 31f, 36f, 41f, 46, 47ff, 84ff, 99
— heat of  93f, 96
— on defect sites  40ff, 94f
— on terrace sites  40ff, 94f
Anharmonic vibrations  75, 81f

Beam-seeding  5

Calibration function  83, 84ff, 92f
Characterization of surfaces during sputtering
— antiphase monitoring  63, 64ff, 67ff
— at high temperatures  63, 67ff
— at low temperatures  34f, 63
— at medium temperatures  63, 64ff
— damaging curve  34, 64
— inphase monitoring  63f, 65
Coadsorption  45ff
2D-condensation  36ff, 46, 95f
Corrugation of the interaction potential  10f, 84f
Cross-section for diffuse scattering
— angular dependence of  20f
— definition of  15
— effective  22f, 45
— measurement of  16, 28, 40ff, 87f
— of adsorbates  9, 13, 16, 20f, 32, 37, 39, 40, 45, 85ff
— of atomic steps  40, 58, 60ff, 64
— of gas phase molecules  16ff, 22, 24
— of vacancies  26, 28, 34ff, 62, 85ff
— overlap of  22f, 29ff, 40ff, 46, 47ff
— velocity dependence of  17ff
Crystal growth  71f

Debye temperature  12, 31f, 84, 86, 91
Desorption experiments  91, 97f
Diffraction peaks (including specular)
— height  10, 14, 58, 61f, 62ff, 73f, 76f, 81f
— profile  14f, 57f, 61f, 74, 77, 78f
— width  57f, 60f, 73, 76, 82
Diffuse scattering *see also* Cross-section for diffuse scattering
— mechanism for  14, 17ff
Diffusion *see* Migration
Dispersion forces  9, 12, 16ff, 34
Dynamical range  6, 11, 13f, 31

Electron Density  11f, 19
Equilibrium experiments  91ff
— isobars  92, 95f
— isosteres  92, 95f
— isotherms  92ff

Facet model  57, 60f

"Ideal" and real surfaces
— characterization  52, 57f
— preparation  7
Interaction between scatterers
— attractive  30, 34ff, 36ff, 40, 46, 62, 96
— repulsive  29f, 32f, 43, 47ff, 74, 94
Interaction potential
— He–adsorbate  12f, 16ff
— He–substrate  9
Interference effects
— reflecting adsorbates  13, 22, 26ff, 84ff
— stepped surfaces  54ff, 57ff, 64ff, 67ff, 71f, 76f
Ion bombardment  34ff, 41f, 60ff, 62ff
Island formation  30f, 36ff, 47ff, 97

Kinetics
— of adsorption  88ff, 98f
— of desorption  97f

107

# inger Tracts in Modern Physics

At the time when this book was made the authors of some of the following volumes of this series were still busy writing their manuscripts, and in other cases the completed typescripts were still in the process of being transformed into finished books. The titles are listed alphabetically by author:

V. Devanathan, A. Nagl, H. Überall
## Nuclear Pion Photoproduction
ISBN 3-540-50671-3


S. Hunklinger, M. Klinger, K. Shvarts
## Physics of Nonmetallic Glasses
ISBN 3-540-50672-1


P. Mulser
## High Power Laser-Matter Interaction
ISBN 3-540-50669-1


V. N. Oraevsky et al:
## Artificial Plasma Clouds
ISBN 3-540-50670-5


A. Hasegawa
## Optical Solitons in Fibers
1989. X, 75 pp. 22 figs. Hardcover  ISBN 3-540-50668-3

Already from the title it is obvious that this volume deals with the most recent developments associated with fashionable and very important theoretical and practical developments: **Solitons** as analytical solutions of nonlinear partial differential equations were established as recent as 1967. Only five years later Hasegawa and Tappert predicted for the first time theoretically that solitons could be generated in a dielectric **fiber.**

The practical implications point towards a technological advance allowing for an economic and undistorted propagation of signals revolutionizing telecommunications.

Starting from an elementary level readily accessible to undergraduates the pioneer in the field provides a clear up-to-date exposition of the prominent features of theoretical background and experimental results in this new and rapidly evolving branch of sciences. This well-written book makes not just easy reading for the researcher but also for interested physicists, mathematicians, and engineers and it is well suited for undergraduate or graduate lecture courses.

er-Verlag Berlin Heidelberg New York London Paris Tokyo Hong Kong

# Springer Tracts in Mode

## Sp

Within this long-established serie
umes on themes which are relate
book and which may be of intere
leagues:

**Volumes 68, 74, 87, 93, 98, 104,
106, 109, 110, 111, 114.**

The subject dealt with in this boo
related to the theme of the follow:

Volume 91

**K. Heinz, K. Müller, T. Engel, K. H**

## Structural Studies

1982. 180 pages. Hard cover. ISBN

"This is an excellent, up-to-date, au
account of two particular approach
surface structure. The book, which
able for those already involved or e
the field of surface science where th
on the application of modern physi

*Journal*

"This is an invaluable reference guid
in the field of atomic and molecular
surfaces (experimental and theoretic

"In summary, this is a worthy additic
Tracts and can be recommended ent

Springer-Verlag Berlin Heidelberg New York London Paris Tokyo Hong

Sprin

S

# Springer Tracts in Modern Physics

At the time when this book was made the authors of some of the following volumes of this series were still busy writing their manuscripts, and in other cases the completed typescripts were still in the process of being transformed into finished books. The titles are listed alphabetically by author:

V. Devanathan, A. Nagl, H. Überall
## Nuclear Pion Photoproduction
ISBN 3-540-50671-3

S. Hunklinger, M. Klinger, K. Shvarts
## Physics of Nonmetallic Glasses
ISBN 3-540-50672-1

P. Mulser
## High Power Laser–Matter Interaction
ISBN 3-540-50669-1

V. N. Oraevsky et al:
## Artificial Plasma Clouds
ISBN 3-540-50670-5

A. Hasegawa
## Optical Solitons in Fibers
1989. X, 75 pp. 22 figs. Hardcover  ISBN 3-540-50668-3

Already from the title it is obvious that this volume deals with the most recent developments associated with fashionable and very important theoretical and practical developments: **Solitons** as analytical solutions of nonlinear partial differential equations were established as recent as 1967. Only five years later Hasegawa and Tappert predicted for the first time theoretically that solitons could be generated in a dielectric **fiber.**

The practical implications point towards a technological advance allowing for an economic and undistorted propagation of signals revolutionizing telecommunications.

Starting from an elementary level readily accessible to undergraduates the pioneer in the field provides a clear up-to-date exposition of the prominent features of theoretical background and experimental results in this new and rapidly evolving branch of sciences. This well-written book makes not just easy reading for the researcher but also for interested physicists, mathematicians, and engineers and it is well suited for undergraduate or graduate lecture courses.

Springer-Verlag Berlin Heidelberg New York London Paris Tokyo Hong Kong

# Springer Tracts in Modern Physics

Within this long-established series there are several volumes on themes which are related to the subject of this book and which may be of interest to you and your colleagues:

**Volumes 68, 74, 87, 93, 98, 104, 106, 109, 110, 111, 114.**

The subject dealt with in this book is particularly closely related to the theme of the following volume:

Volume 91

**K. Heinz, K. Müller, T. Engel, K. H. Rieder**

## Structural Studies of Surfaces

1982. 180 pages. Hard cover. ISBN 3-540-10964-1

"This is an excellent, up-to-date, authoritative and balanced account of two particular approaches to the elucidation of surface structure. The book, which is well produced, is suitable for those already involved or embarking on research in the field of surface science where the emphasis is mainly on the application of modern physical techniques."

*Journal of the Chemical Society,*
*Faraday Transactions*

"This is an invaluable reference guide for anyone working in the field of atomic and molecular beam scattering at surfaces (experimental and theoretical)." *J. Am. Chem. Soc.*

"In summary, this is a worthy addition to the Springer Tracts and can be recommended enthusiastically."

*Journal de Physique*

Springer-Verlag Berlin Heidelberg New York London Paris Tokyo Hong Kong

Springer